ISSUES IN UNDERGROUND STORAGE TANK MANAGEMENT

TANK CLOSURE AND FINANCIAL ASSURANCE

JANET E. ROBINSON
PAUL THOMPSON
W. DAVID CONN
LEON GEYER

CRC Press
Taylor & Francis Group
Boca Raton London New York

CRC Press is an imprint of the
Taylor & Francis Group, an **informa** business

CRC Press
Taylor & Francis Group
6000 Broken Sound Parkway NW, Suite 300
Boca Raton, FL 33487-2742

First issued in paperback 2020

ISBN 13: 978-0-367-57990-6 (pbk)
ISBN 13: 978-0-87371-402-0 (hbk)

Visit the Taylor & Francis Web site at
http://www.taylorandfrancis.com

and the CRC Press Web site at
http://www.crcpress.com

Library of Congress Cataloging-in-Publication Data

Issues in underground storage tank management: tank closure and financial
    assurance/Janet E. Robinson . . . [et al.].
        p. cm.
    Includes bibliographical references and index.
    ISBN 0-87371-402-4
    1. Hazardous substances—Safety measures. 2. Chemical spills—
Environmental aspects. 3. Storage tanks. I. Robinson, Janet E.
TD1050.S24U53   1993
604.7—dc20   92-36719
                                                                CIP

To my parents and E.S.N.

Janet E. Robinson

To my parents and Bryce, Scott, and Jan

Leon Geyer

Preface

Among the successive waves of environmental issues, underground storage tank (UST) management is unique in several ways. Unlike hazardous waste disposal sites, USTs are ubiquitous in both urban and rural areas due to their use for storing heating oil and gasoline, arguably the two fuels used most frequently by vehicle owners in North America. UST management is not, unlike other hazardous waste issues, associated primarily with industrial concerns, but rather with a wide spectrum of (mostly) smaller entities: private citizens, service stations, local governments, schools, and almost everyone else storing significant quantities of fuel on site. As the result of sweeping federal regulations promulgated in the late 1980s, many individuals who had been involved only marginally in environmental protection programs, now find themselves pouring through EPA documents and consulting with engineers to remove their tanks in compliance with federal or state requirements. In addition to tank closure, most UST owners and operators face the still-unresolved dilemma of providing financial assurance against UST leaks and spills when insurance is either unavailable or prohibitively expensive. Besides the hundreds of thousands of people who must comply with these regulations, every citizen has an interest in seeing that the UST closure and financial assurance regulations protect the environment and public health in an efficient and equitable fashion.

Although closure and financial assurance are critical elements of the public's response to leaking underground storage tanks, detailed information on the options and issues associated with these two topics remains in short supply. We developed this book to serve as an information source for tank owners, consultants, interested citizens, policy analysts, and decision-makers who want to understand the options and strategies available to close tank systems and meet federal financial assurance requirements. The second part of the book which focuses on financial assurance requirements, also assesses the successes and failures of the decision-making processes that have led to the current regulations. The authors — professionals in the fields of engineering, planning, and law — hope this book appeals to a wide variety of interests. Our goal in writing it was to assist those interested in UST management — from both the public and private sectors — in making well-informed UST decisions, be they economic, technical, or political.

The Authors

Janet E. Robinson is a project scientist with Woodard & Curran Environmental Services, Inc. in Portland, Maine. She has been involved with hazardous waste and underground storage tank issues since 1985, when she developed and wrote a series of technical support documents that were widely distributed to state and local policymakers in Virginia. As a consultant, her work is concerned primarily with UST management, facility decontamination and closure, remedial investigation, and environmental auditing. She has authored over 10 publications on a wide variety of hazardous waste issues.

Ms. Robinson holds a Bachelor of Science degree from Virginia Polytechnic Institute and State University (VPI & SU) and a Master's degree from VPI & SU and Marshall University, West Virginia.

Paul Thompson is a policy analyst with the Center for Policy Research of the National Governors' Association (NGA). Mr. Thompson's work has focused primarily on evaluating the role of the states in federal solid and hazardous waste management programs, including those established pursuant to the Resource Conservation and Recovery Act and the Comprehensive Environmental Response, Compensation, and Liability Act. Mr. Thompson also has worked for the Natural Resources Defense Council, specializing in the analysis of state and local programs to reduce nonpoint source water pollution under the Clean Water Act.

Mr. Thompson has a Master's degree in Urban and Regional Planning from Virginia Polytechnic Institute and State University, where his research focused on the local infrastructure for hazardous waste management programs, the impact of federal underground storage tank regulations on small petroleum marketing firms, and the development of state capacity to comprehensively manage groundwater resources. Mr. Thompson also has a Bachelor of Arts degree in Political Science and a Bachelor of Science degree in Biology from Roanoke College.

A native of Virginia, Mr. Thompson currently lives in Baltimore, Maryland, with his wife Ellen.

W. David Conn is a professor and the associate director of the University Center for Environmental and Hazardous Materials Studies at Virginia Polytechnic Institute and State University. He also heads the Ph.D. Program in Environmental Design and Planning.

Dr. Conn has degrees in both chemistry and economics from Oxford University, and taught environmental policy/planning for eight years at U.C.L.A. before coming to VPI & SU in 1980. He spent the 1991–92 academic year as an American Council on Education Fellow at the University of Tennessee,

Knoxville. He has had over twenty-two years of experience in research, professional application, and instruction with a primary emphasis on evaluation, risk management (including risk communication), and pollution prevention, largely related to solid/hazardous wastes and other hazardous substances. In 1986–87 he was co-director of a major study on the management of underground storage tanks, funded by the Virginia General Assembly and the U.S. Geological Survey through the Virginia Water Resources Research Center.

In addition to co-authoring the present volume, Dr. Conn has published and edited books as well as many chapters and articles for a variety of scholarly and professional publications.

L. Leon Geyer is an associate professor of law and economics at Virginia Polytechnic Institute and State University. An associate of the Center for Environmental and Hazardous Materials Studies, he teaches environmental law, real estate law and appraisal, agricultural law, and food and agricultural policy. Professor Geyer received his Ph.D. from the University of Minnesota and his J.D. from Notre Dame. He has been the Altheimer Visiting Professor of Law, University of Arkansas at Little Rock, and a former attorney for the House of Representatives Committee on Agriculture and Office of General Counsel, U.S.D.A. Professor Geyer has received numerous awards for teaching, is a member of the bars of Indiana, District of Columbia, and Supreme Court of the United States. He has published over 50 professional papers and given over 130 state, national, and international professional presentations. He currently serves as Faculty Senate President.

Acknowledgments

The authors would like to acknowledge the assistance of the following organizations who provided information or support for this work:

Woodard & Curran Environmental Services Inc., Portland, Maine
Clean Harbors of Maine, Inc., South Portland, Maine
Jetline Services, Portland, Maine
Dup's Inc., Houston, Texas
Environmental Products and Services, Inc., Buffalo, New York
Massachusetts Tank Disposal Inc., Chicopee, Massachusetts
Laidlaw Environmental Services, Pinewood, South Carolina
Chemical Waste Management, Emelle, Alabama
American Petroleum Institute, Washington, D.C.
U.S. Environmental Protection Agency, Region 5, Chicago, Illinois
PEI Associates, Washington, D.C.
New York Dept. of Environmental Conservation, Region 9
A & A Environmental Services, Baltimore, Maryland
Tank Tech Corp., Congers, New York
Environmental Tank Testing Inc., Baltimore, Maryland
The Planning Corporation, Reston, Virginia
Front Royal Insurance Co., McClean, Virginia
Petroleum Marketers Association of America, Washington, D.C.
Collier, Shannon, Rill and Scott, Washington, D.C.
Washington Pollution Liability Insurance Agency, Olympia, Washington
Sedgwick James of Pennsylvania, Inc., Harrisburg, Pennsylvania
U.S. Environmental Protection Agency, Office of Underground Storage Tanks, Washington, D.C.

The analyses and opinions expressed by Mr. Thompson are his own and do not reflect the policies or opinions of his employer.

Contents

List of Tables

List of Appendices

List of Acronyms

API	American Petroleum Institute
BTEX	benzene, toluene, ethylbenzene, and xylene
BOCA	Building Officials and Code Administration
CERCLA	Comprehensive Environmental Response, Compensation, and Liability Act
CFR	Code of Federal Regulations
CGI	combustible gas indicator
DOT	Department of Transportation
EDB	ethylene dibromide
EM	electromagnetic conductivity
EPA	Environmental Protection Agency
FID	flame ionization detector
FR	Federal Register
GC	gas chromatograph
GPR	ground penetrating radar
HSWA	Hazardous and Solid Waste Amendments (to RCRA) of 1986
LEL	lower explosion limit
PID	photoionization detector
MCL	maximum contaminant level
MS	mass spectrometer
MTBE	methyl-tertiary-butyl-ether
NFPA	National Fire Protection Association
NIOSH	National Institute of Occupational Safety and Health
NPL	National Priority List
NRC	National Response Center
OSHA	Occupational Safety and Health Administration
OVA	organic vapor analyzer
OVM	organic vapor monitor
RCRA	Resource Conservation and Recovery Act
SARA	Superfund Amendments and Reauthorization Act
TCLP	Toxicity characteristic leaching procedure
TPH	total petroleum hydrocarbons
TSD	treatment, storage, and disposal
UST	underground storage tank

PART I
UNDERGROUND STORAGE TANK CLOSURE

Introduction

The closure of old and unwanted underground storage tanks (USTs), formerly handled in a relatively straightforward if often somewhat environmentally questionable manner, has recently become more problematic as both the quantity of tanks removed from service and the regulations affecting the practice increase in number. Evolving federal and state regulations that contain strict technical and performance standards are resulting in the closure of thousands of unneeded or substandard USTs each year. Although tank management (as opposed to closure) is governed largely by a single set of state or federal regulations, at the time of closure both hazardous waste laws and the strict liability mandates of the "Superfund" program come into play and affect both the availability and the environmental desirability of established disposal options. To many, it seems that the problem of final disposal gets bigger while the solutions become fewer; this view, while not entirely accurate, does illustrate the need for tank owners to have a thorough understanding of all the applicable environmental laws and the technologies and strategies realistically available to deal with them, before embarking on costly tank removal and replacement programs.

This section of the book describes various technical, economic, and environmental issues pertaining to underground storage tank removal and disposal and is intended as a technical guide for those directly or indirectly involved in tank management: tank owners, environmental managers, government regulators, commercial contractors and consultants, and other interested parties. While the emphasis is primarily on tanks used to store petroleum and "other regulated substances" (i.e., those subject to regulation under Subtitle I of the Resource Conservation and Recovery Act, 40 CFR 280 and 281), many of the issues and considerations are the same for hazardous waste tanks.

For the tank owner, this section on tank closure provides the information necessary to select a tank closure technology and strategy that are both economically and environmentally sound, and to manage or subcontract tank closure services efficiently. Special consideration is paid to complying with hazardous waste laws and to minimizing the longterm risk associated with hazardous waste disposal. Guidelines for selecting disposal technologies and strategies are presented in relation to current hazardous waste regulations, for both the tank and its contents.

A detailed section is included on site assessment, a relatively new component of traditional tank closure, so that the tank owner can be an informed participant in site investigation activities. The chapter should also prove useful to consultants and others involved in assessing contamination at underground storage tank sites. Similarly, Chapter 6 discusses site health and safety, a vital concern to everyone on the site.

The information presented in the following chapters was drawn from a number of sources: published literature; interviews with government representatives, tank removal contractors, and other involved parties; and the author's own experience as an environmental consultant and researcher.[1,2] Significant components of the tank closure field are represented through this collective voice, but because of each job's variability in site conditions and applicable regulations, recommendations and advice are sometimes necessarily general. Virtually all tank closures, however, have important elements in common, and those common issues are the focus of this half of the book.

REFERENCES

1. Robinson, J. E. et al., *Underground Storage Tank Disposal: Procedures, Economics, and Environmental Costs*, Virginia Water Resource Research Institute Bull. No. 161, Blacksburg, Virginia, 1988.
2. Robinson, J. E. et al. *Underground Storage Tank Management: A Guide for Local Governments*, Virginia Polytechnic Institute and State University Hazardous Waste Management Project, Blacksburg, Virginia, 1986.

CHAPTER 1

Tank Closure, Step-By-Step

Janet E. Robinson

CONTENTS

0-87371-402-4/93/$0.00+$.50

Tank Closure, Step-By-Step

Tank closure is characterized by a series of sequential decisions, each requiring knowledge of both options and consequences so as to avoid the cost and delays potentially associated with misjudgment. The purpose of this book is to provide the information necessary to make informed and effective choices; the purpose of this chapter is to present an overview of the closure process, highlighting particular issues and sequences that need to be resolved or arranged early in the closure effort. Some of these issues will be discussed in more detail in later chapters.

In many cases, tank removal is viewed and treated as a precursor to a larger, more "important" effort, such as the sale of a property or the installation of a new tank system. As a side-service formerly performed by general contractors, tank removal is not always perceived to be a specialized task requiring technical expertise and careful planning. Although the removal of many tanks, especially small ones, is sometimes accomplished in a relatively straightforward manner, tank owners who forego adequate planning and preparation risk incurring unexpected delays and costs when permits must be acquired at the last minute, when premium laboratory fees are charged for accelerated turnaround times, and when delays in paperwork prohibit removal of drummed tank waste from the site.

For these reasons, tank removal should be approached as the potentially complex, potentially lengthy, and expensive task that it sometimes is. By knowing what the job involves, what the options are, who should do the work, and how long it could take, the tank owner can allocate sufficient time and money so that important contingent actions, such as loan approvals or property sales, can take place on schedule.

1.1 SEQUENCE OF EVENTS

In order for closure to proceed smoothly and in accord with applicable requirements, the following key points should be considered at the outset.

1.1.1 Select Closure Timeframe

Although federal and many state regulations set specific schedules for tank removal and upgrading, in some cases a tank owner may find economic advantage by moving in advance of the regulations and removing all the facility's tanks at once. While incurring a large single expense, early removal nearly eliminates future spill liability, and avoids the need to comply with additional future regulations or guidelines that may result from evolving state underground storage tank (UST) programs. In addition, since the open holes associated with tank removal often translate into decreased business for retail motor fuel operators, tearing up the site only once may reduce the overall loss in sales. This is particularly true if the work is scheduled for a period such as the off-season in tourist areas, when sales are normally low.

1.1.2 Learn the Regulations

As described in more detail in Chapter 2, underground storage tanks may be governed by federal, state, and local regulations. Federal laws set a minimum standard, but since many state or local laws exceed the federal ones in applicability or requirements, tank owners need to find out exactly what the regulations are in their area and what steps they need to follow to comply with them. These regulations will, in turn, dictate arrangements that should be made before closure actually takes place.

Before fieldwork begins, the following regulatory issues should be investigated.

- **Permits** — Tank owners should investigate what permits are required to close a tank in the area. Permits may be required from the building inspector, the fire marshall, the county health department or environmental agency, or the state UST regulatory agency. A phone call to each of these offices will get the paperwork started; some contractors will complete much of these requirements as part of their tank closure services. No permit is currently required under federal regulations.
- **EPA/State Involvement** — Some states and local regulations require that an agency representative be on hand to observe or inspect UST closures, especially those initiated in response to a suspected leak or when signs of leakage are discovered during excavation. Because local officials or state representatives usually have enormous workloads, their availability may be limited and require that they be notified in advance.
- **Waste Disposal** — Tank sludge and waste product that meets the federal or state definition of "hazardous waste" will require the tank owner to acquire an EPA Identification (ID) Number before transporting or disposing of the material. In some states a Temporary Generator ID Number can be obtained over the phone, but in others a written form may be required.

1.1.3 Remove, Repair, or Retrofit

Federal regulations provide three options for existing underground tanks: removal, abandonment in place, or upgrading (retrofitting) to meet new tank standards. Removal is usually the most widely chosen alternative, especially since abandonment in place is frequently prohibited or restricted by local regulations. Retrofitting has certain short-term advantages, especially for relatively new tanks or for those which will not be needed after 1998, the federal deadline for upgrading tanks. It is important to remember that "closure" is not necessarily synonymous with "removal", although it usually is; the available alternatives should be considered carefully before closure is initiated. These alternatives are discussed in more detail in Chapter 3.

1.1.4 Construct Closure Timeline

When tank closure must be completed by a given date, all the steps in the closure process must be taken into account so that the preparation and field-work can begin early enough to ensure meeting the deadline. In addition to the regulatory issues described above, adequate time should be allowed to complete the steps described below.

- **Contractor Selection** — Because improperly abandoned or installed underground tanks have been the cause of numerous environmental problems in the past, some states have implemented certification programs for tank installers. While increasing the quality of tank work, certification requirements may mean that fewer contractors are available than in previous years, when many general contractors performed tank removals. Thus, locating and securing the services of a qualified contractor may take longer than in the past. Advice on contractor selection is contained in Chapter 5.
- **Laboratory Analysis** — Normal turnaround time for many analytical laboratories is 30 days. "Rush" jobs, completed in less than the normal period, demand premium rates, which are often 150 to 200% of the regular fee. Factoring adequate analytical time into the closure schedule can save a tank owner hundreds of dollars in laboratory fees.

A schedule of closure is shown in Table 1.

Table 1. Recommended Sequence of Tasks in Tank Closure

1. Select closure timeframe and strategy
2. Review federal, state, and local closure requirements
3. Select and schedule closure contractor(s)
4. Notify authorities, if necessary
5. Obtain EPA Identification Number, if product or sludge is present in tank
6. Perform closure and site assessment
7. Receive analytical results, if any
8. Notify authorities of closure completion, if required
9. Safely file all records and results concerned with tank closure

CHAPTER 2

The Law

Janet E. Robinson

CONTENTS

0-87371-402-4/93/$0.00+$.50
© 1993 by Lewis Publishers

The Law

2.1 LAWS AND REGULATIONS: INTRODUCTION

Because of the heightened public impact of groundwater contamination problems and the resulting regulatory attention of governments at all levels, any underground storage tank (UST) may potentially be governed by local, state, and federal regulations. Federal regulations, published on September 23, 1988, establish minimum standards for a carefully defined universe of underground tanks; but state and local governments, in order to respond to the specific needs of their areas, may impose additional requirements on a still wider population of tanks (including some not yet subject to federal law). In addition, state and local requirements may be applied in ways different from federal law, for instance, by means of a permit system for tanks rather than a general timetable with deadlines for upgrading tanks. A thorough knowledge of applicable laws and regulations is essential to avoid the delays, expenses, and legal liabilities that are the nemesis of so many UST removals; this section therefore provides a summary of the various regulations or regulatory vehicles which have relevance to UST management in general.

These various laws, and the ways in which they pertain or may pertain to UST management are presented in Table 2.

In this chapter, each level of regulation will be discussed as it pertains to tank closure.

A note on legal language, as cited in the text: the intentions of Congress, as embodied in federal environmental Acts, are carried out through specific regulations devised, promulgated, and enforced by EPA. The Act is usually known by an acronym, such as RCRA or CERCLA. The regulations published to implement the Act are contained in the Code of Federal Regulations (CFR), which is divided into numbered Parts; thus 40 CFR 280 refers to volume 40 of the Code of Federal Regulations, Part 280. Titles and alphabetical designations are used to describe topically various sections of regulations; e.g., 40 CFR 280.70 to 280.74 is also entitled "Subpart G: Out-of-Service UST Systems and Closure".

Additions to or modifications of the CFR are introduced in the Federal Register, a daily publication of the federal government's regulatory activity. The Federal Register presents not only the legal text of the regulation but, in the preamble preceding it, an informative and usually fairly readable discussion of the regulation and EPA's reasons for promulgating it. The Federal Register is recommended reading for anyone wanting a more thorough

Table 2.   Applicable or Potentially Applicable Laws and Regulations

Act or Regulation	UST-Related Contents
Federal	
RCRA Subtitle I	UST technical standards and corrective action requirements; obtain information from the state agency or the regional EPA office (see Appendices 1 and 2)
RCRA Subtitle C	Hazardous waste regulations; obtain information from the state agency or regional EPA (see Appendices 1 and 2)
CERCLA ("Superfund")	Owner/operator liabilities, release reporting requirements, and cleanup requirements
State	
State UST programs	State UST regulations, either the same as or more stringent than federal regulations
State building codes	Statewide construction regulations, which may contain UST removal or abandonment criteria and notification or inspection requirements
Local	
Fire codes	UST removal or abandonment criteria, and notification or inspection requirements; in some areas the Fire Department may be the implementing agency for federal regulations; see Appendices 1 and 2
Local building codes	Local building codes with UST removal or abandonment criteria and notification and inspection requirements; these may be different from state codes; obtain information from the local building inspector
Zoning laws	Removal or abandonment criteria; tank location restrictions; obtain information from the local zoning board or town planner

understanding of the content and intent of federal requirements. Federal Register entries are cited formally by volume and page number, followed sometimes by the date of issue: e.g., 53 Federal Register 37082 means Volume 53 of the Federal Register page 37,082, which was published on September 23, 1988.

Both the Code of Federal Regulations and the Federal Register are available in law libraries (maintained by some county courthouses) and many university and public libraries.

2.2 FEDERAL LAWS

2.2.1 The Resource Conservation and Recovery Act (RCRA)

The Resource Conservation and Recovery Act (RCRA) of 1976 is the major federal vehicle for hazardous waste and underground storage tank regulation. Regulations promulgated under Subtitle I of RCRA specify UST standards and requirements, while those under Subtitle C of RCRA govern the generation, transportation, storage, treatment, and disposal of hazardous waste. Since the liquid waste remaining in tanks may meet the federal definition of a "hazardous waste", tank owners should be familiar with regulations published under both parts of the Act to be able to make intelligent choices about disposal options.

2.2.1.1 RCRA Subtitle I: Underground Storage Tanks

Part 280 of the RCRA regulations, entitled *Technical Standards and Corrective Action Requirements for Owners and Operators of Underground Storage Tanks,* contains the federal requirements for upgrading schedules, for closure and corrective action, and for performance standards for new and existing USTs. Copies of the regulations are obtainable from the regional EPA offices; however, state regulations may be more stringent than these.

2.2.1.1.1 *Key Definitions under RCRA Subtitle I*

Several terms used in the RCRA UST regulations have specific meanings which may be different from their popular definitions. Selected key terms, and their federal definitions as summarized from 40 CFR 280.12, are provided below.

Connected piping — "All underground piping, including valves, elbows, joints, flanges, and flexible connectors attached to a tank system through which regulated substances flow. For the purpose of determining how much piping is connected to any individual UST system, the piping that joins two UST systems should be allocated equally between them."

Empty — An UST is empty if: "(i) All wastes have been removed that can be removed using the practices commonly employed to remove materials... *and* (ii) No more than 2.5 centimeters (one inch) of residue remain on the bottom..., *or* (iii)(A) No more than 3% by weight of the total capacity of the container remains in the container or inner liner if the container is less than or equal to 110 gallons in size, or (B) No more than 0.3% by weight of the total capacity of the container remains in the container or inner liner if the container is greater than 110 gallons in size" [53 FR 37182; 40 CFR 261.7(b)(1)].

Existing tank system — "A tank system used to contain an accumulation of a regulated substance of or for which installation commenced on or before December 22, 1988."

New tank system — "A tank system that contains a regulated substance and for which installation commenced after December 22, 1988."

Out-of-service USTs — These are not specifically defined by the federal regulations. For some tanks, such as emergency generator and backup fuel system tanks, the normal use pattern involves long periods of disuse; EPA does not consider these tanks "out-of-service" and thus subject to mandatory closure requirements. Guidance is provided in the preamble to the final UST regulations (53 FR 37182):

> "In determining whether a tank is in or out of service, the tank owner should consider the following:
>
> * Whether the tank is included in the normal operation and maintenance procedures at the facility
> * The types and amounts of regulated substances stored at the facility
> * The likelihood that an undetected leak has occurred or may occur in the future, and
> * The potential that the tank has become a   receptacle for illegal dumping
>
> If, however, the infrequent use of a tank cannot be justified as part of its purpose and/or if the operation, maintenance, or release detection procedures associated with the tank are inadequate or inconsistent with the monitoring procedures required for operating tanks under federal or state regulations, the tank will be considered out of service, and hence, temporarily closed."

Out-of-service tanks are subject to the temporary and permanent closure requirements discussed in the next section.

Owner — "(a) In the case of an UST system in use on November 8, 1984 or brought into use after that date, any person who owns an UST system used for the storage, use, or dispensing of regulated substances; and (b) in the case of any UST system in use before November 8, 1984, but no longer in use on that date, any person who owned such UST immediately before the discontinuation of its use."

Person — "An individual, trust, firm, joint stock company, Federal agency, corporation, state, municipality, commission, political subdivision of a state, or any interstate body. 'Person' also includes a consortium, a joint venture, a commercial entity, and the United States Government."

Tank — "a stationary device designed to contain an accumulation of regulated substances and constructed of non-earthen materials that supply structural support." This definition has been interpreted as excluding some in-ground concrete tanks, for which the surrounding earth provides structural support.[1]

Underground storage tank — "Any one or combination of tanks (including underground pipes connected thereto) that is used to contain an accumulation of regulated substances, and the volume of which (including the volume of underground pipes connected thereto) is 10 percent or more beneath the surface of the ground." (See Table 3 for exclusions.)

Table 3. Underground Storage Tanks Excluded from Federal Regulation under RCRA Subtitle I (40 CFR 280.12)

Farm or residential tanks of 1100 gallons or less capacity storing motor fuel for noncommercial purposes

Tanks storing heating oil for consumptive use on the premises where stored

Septic tanks

Pipeline facilities (including gathering lines) regulated under the Natural Gas Pipeline Safety Act of 1968, the Hazardous Liquid Pipeline Act of 1979, or State laws comparable to these Acts

Surface impoundments, pits, ponds, and lagoons

Stormwater or wastewater collection systems

Flow-through process tanks

Liquid traps or associated gathering lines directly related to oil or gas production and gathering operations

Storage tanks situated on or above the floor of underground areas, such as basements and cellars.

UST system or tank system — "An underground storage tank, connected underground piping, underground ancillary equipment, and containment system, if any."

Although federal UST regulations apply to a wide variety of subsurface structures, several important exclusions apply. The first task of a tank owner is to determine whether the tanks in question fall within the federal definition of regulated units, as described below.

2.2.1.1.2 Units Regulated by RCRA Subtitle I

Under federal law, "underground storage tank" means any one or combination of tanks (including piping) whose total volume is more than 10% underground. The regulations apply to any tank used to store petroleum products or a "hazardous substance" as defined by CERCLA (the Superfund Law; see Section 2.2.3).

These regulations do not apply to tanks containing a RCRA-defined "hazardous waste"; hazardous waste tanks are regulated separately under Subpart J of RCRA (40 CFR 264.190 to 264.199 or 265.190 to 265.199). In addition, numerous kinds of units are excluded from the universe of subsurface units regulated by RCRA Subtitle I because they do not fit the federal definition of "tank" described above. Other federal exclusions are shown in Table 3. Important to note, however, is that tanks excluded from federal regulation may still be covered by state or local regulations; again, federal regulation represents *minimum* standards only.

Technical and operating standards for new and existing tank systems comprise the bulk of these UST regulations, and are explained in numerous technical articles and publications.

Federal closure regulations differ according to whether the tank is temporarily closed or permanently closed. Requirements associated with each category are discussed below.

Table 4. Summary of Federal Requirements for Temporarily-Closed Tanks

Time Closed	Technical Requirements
0–3 months	Operate and maintain corrosion protection and (unless the tank is empty) leak detection
3 or more months	Operate and maintain corrosion protection and (unless the tank is empty) leak detection
	Open vents; cap and secure all other lines, pumps, entryways, and ancillary equipment
12 or more months	Meet new tank standards for corrosion protection of tanks and piping, with the exception of spill and overfill equipment requirements;
	or
	Comply with the schedule and requirements for upgrading tanks, with the exception of spill and overfill requirements;
	or
	Close the tank;
	or
	Apply for an extension of the 12-month temporary closure period

Note: In all options above, release detection procedures may be discontinued if the tank is "empty" as defined by RCRA. Any releases detected during temporary closure must be responded to in accordance with 40 CFR Subparts E and F.

2.2.1.2 Temporary Closure (40 CFR 280.70)

Under federal regulations, tanks may be removed from service for up to 12 months without requiring permanent closure, regardless of whether they still contain product or not. Requirements for temporary closure vary somewhat with the length of time that the tank is out of service, and consist of many of the same technical standards (corrosion protection, release detection, etc.) applicable to in-service tanks. Temporary closure requirements are described below and summarized in Table 4.

All tanks which are temporarily closed must continue to meet the same corrosion protection and release response requirements applicable to operating tanks. However, if the tank is "empty" (see Section 2.2.1.1.1), release detection is not required. Likewise, since temporarily closed tanks would not have problems with spilling and overfilling, temporarily closed UST systems are not required to satisfy the spill and overfill requirements for new and upgraded systems in order to be excluded from the permanent closure requirements that are normally imposed after 12 months (53 CFR 371.83). In the event that a leak from the tank is suspected or confirmed, release investigation and corrective actions procedures must be implemented in accordance with Subtitle E and F of federal UST regulations.

When an UST system is closed for 3 months or more, all lines, pumps, access holes, and ancillary equipment must be capped and secured. Vent lines must be left open and functioning, and the corrosion protection and release response activities (if implemented) must continue.

When an UST is closed for more than 12 months, tank owners must permanently close the UST system unless either of the following apply [40 CFR 280.70(c)]:

1. The tank meets the technical performance standards for new tanks (as described in 40 CFR 280.20), specifically, leak detection and corrosion protection;
2. The tank owner complies with the schedule and requirements for upgrading existing tanks, as defined in 40 CFR 280.21. Under this section, tank owners must install leak detection according to EPA's schedule, and add corrosion protection by December of 1998.

As noted above, spill and overfill prevention requirements, normally a part of new and upgraded tank standards, do not have to be met by temporarily closed UST systems.

Any temporarily closed UST system that does not comply with either the new tank performance standard or the upgrade requirements for existing tanks must permanently close after the 12-month temporary closure period ends (53 FR 37183). Tanks that are in compliance, however, may remain closed indefinitely, and need not be subject to any special integrity-testing procedures prior to being reopened.

Also, existing temporarily closed tanks which are empty need only comply with the upgrading requirements for corrosion protection, which do not go into effect until December 1998. Thus, an empty, existing tank (i.e., one installed prior to December 22, 1988) can remain "temporarily closed" until December 22, 1998, provided that it remains empty. If a regulated substance is added to the tank, however, the tank must be equipped with an approved release detection device, according to the implementation schedule of 40 CFR 280.40.

In some situations where a tank may need to remain out of service for more than 12 months before being reactivated, the regulating state agency or EPA can provide a special extension of the temporary closure period to prevent these tanks from being closed unnecessarily. Tank owners must complete a site assessment conducted according to 40 CFR 280.72 (described in Section 2.2.1.3.2) before requesting a special extension.

2.2.1.3 Permanent Closure and Changes in Service (40 CFR 280.71)

2.2.1.3.1. Components

According to federal requirements, an UST may be "closed" in one of three ways: (1) removal, (2) abandonment in place, or (3) change in service, that is, altering its use to that of storing a nonregulated substance.

To close a tank permanently, a tank owner must perform the following general steps:

1. Contact the state agency or EPA at least 30 days before beginning site closure (unless the closure is in response to a corrective action).
2. Empty the tank of all stored material. A tank is considered "empty" when less than 1 inch or 0.3% by weight of the total tank system capacity is left in the tank.
3. Clean the tank by removing all accumulated sludges and residues.
4. "Close" the tank by: (1) removing it from the ground; (2) filling it with an inert, solid material; or (3) changing its use to that of storing a nonregulated substance.
5. Conduct a site assessment (40 CFR 280.72) before completing closure *unless* a release detection method, in place at the time of closure, indicates that no release has occurred.

The specific procedures to be followed in tank removal or abandonment should conform to industry-recommended practices, as outlined in the following standards published by the American Petroleum Institute (API):

- API 1604: "Removal and Disposal of Used Underground Petroleum Storage Tanks"
- API 2015: "Cleaning Petroleum Storage Tanks"
- API 1631: "Interior Lining of Underground Storage Tanks"

For some hazardous substance tanks, the regulations also specify the use of the following documents and guidance:

- "Criteria for a Recommended Standard: Working in a Confined Space," by the National Institute for Occupational Safety and Health (NIOSH).

Although these documents are specified as guidance by the final UST regulations, a variety of other organizations publish related standards, such as the following two useful publications from the National Fire Protection Association (NFPA):

- NFPA 30: "Flammable and Combustible Liquids Code, 1987 Edition"
- NFPA 329: "Underground Leakage of Flammable and Combustible Liquids, 1987"

Like other technical documents, these standards are upgraded periodically to keep pace with new knowledge and technical developments. However, tank owners must only adhere to the versions in effect in December, 1988, the time at which the final UST regulations were promulgated, regardless of subsequent revisions. Since EPA considers those versions sufficiently comprehensive to achieve the level of environmental and human health protection intended by the regulations, tank contractors and others need not endeavor each year to obtain revised editions.

2.2.1.3.2 *Site Assessment (40 CFR 280.72)*

At some point before the closure of the tank is completed (either before or after removal), the site must be evaluated for the presence of released product *unless* the tank is equipped with an approved external release detection device that indicates that the site is free of released product. Site assessment usually involves the collection of soil and (if necessary) groundwater samples from in and around the tank excavation, in compliance with sound scientific practice or state regulations; federal regulations, however, do not specifically require sampling. What the federal regulations do require is that the selection of sample types, sample locations, and methods of measurement take into account the method of closure, the nature of the stored material, the type of backfill, the depth to groundwater, and other factors necessary to determine the presence of a release (53 FR 37184).

The "approved" external release detection methods consist of either a soil vapor or groundwater monitoring system. These systems must be designed and operated according to the standards of 40 CFR 280.43 in the UST regulations.

If contaminated soil, groundwater, or free product as a liquid or vapor is discovered around the tank, tank owners must notify state UST officials or EPA and begin corrective actions in accordance with Subpart F of the regulations. The selection and implementation of corrective action procedures is discussed by EPA and other authors.[2,3]

In cases where the state agency or EPA believes that a tank closed before the final UST regulation took effect (December 22, 1988) presents a current or potential threat to human health and the environment, the agency may require the tank owner to conduct a site investigation of the evacuation zone and reclose the tank according to the current regulations. According to EPA, more than 300 release incidents reported between 1970 and 1984 were associated with abandoned UST systems, and EPA believes that many tanks improperly closed before the standards promulgated in 1988 may still present a significant threat to human health and the environment (53 FR 37185). Federal law provides EPA or the State UST agency with the authority to require that old tanks or tank locations be investigated, and if necessary, remediated, regardless of when the tank was closed.

2.2.1.3.3 *Closure Records (40 CFR 280.74)*

Tank owners must retain records to show compliance with the closure regulations for a period of at least 3 years after closure or change-in-service is completed. Records should be kept at the site or at a readily available alternative so that they may be inspected on request by the state agency or EPA. The records may be maintained by: (1) the tank owners or operators who took the tank out of service; (2) the current owners or operators of the tank; or (3) the state agency or EPA, to whom the records should be mailed if they cannot be maintained at the closed facility.

2.2.2 RCRA Subtitle C: Hazardous Waste Management

Subtitle C of RCRA delineates a strict set of regulations governing the storage, transport, treatment, and disposal of hazardous wastes. Since the liquid residue sometimes present in the bottom of fuel and hazardous substance tanks often is, or is assumed to be, hazardous according to the RCRA definitions, tank owners and disposers must consequently make sure that their activities, and those performed by contractors or subcontractors, are in compliance with the procedures outlined in Subtitle C of the Act.

2.2.2.1 Manifests and Permits

These regulations apply when over 100 kg (220 lbs, 26 gallons, or about half of a 55-gal drum) of hazardous wastes are produced within a calendar month; generators of less than that are classified as "small quantity generators" and excluded from federal regulation.

The RCRA strategy focuses on two main techniques to ensure safe waste handling: a manifest system to track waste from generator to final disposer ("cradle-to-grave") and a set of permits and standards which apply to all those who generate, store, transport, treat, or dispose of hazardous waste.

A "hazardous waste" is defined as any waste material that (1) exhibits specified characteristics of ignitability, reactivity, toxicity, or corrosivity, or (2) is explicitly identified and numbered by EPA on a series of lists based on industrial source or specific chemical type. According to one of these lists, for instance, the bottoms of leaded gasoline tanks used by the petroleum refining industry are hazardous; this listing does not, however, apply specifically to retail outlets. While tank residues are excluded from current toxicity standards, liquid petroleum products are usually hazardous because of their ignitability.

The manifest system is a strategy to provide a verifiable paper track of a hazardous waste from the time it is generated to its point of final disposal. A "manifest form," which identifies the type and quantity of waste and all those who will subsequently handle it, is filled out be the generator and signed by both the transporter(s) and treatment facility as they handle the waste, with the completed original being returned to the generator ("round-trip" manifesting) as verification of proper disposal. When a tank owner hires a professional tank cleaning and disposal firm, that firm often handles the manifesting paperwork, although the generator must still sign the manifest form.

2.2.2.2 Generator Responsibilities

RCRA Subtitle C regulations impose a number of specific responsibilities on the generators of hazardous waste. To dispose of hazardous waste, the tank owner or any contracted agent who handles a potentially hazardous waste must do the following:

- Determine whether the waste is hazardous;
- Obtain an EPA Identification number;
- Restrict on-site storage of the waste to no more than 180 days, or 270 days if it must be transported more than 200 miles for treatment (if more than 220 lbs of waste are involved, the waste must be shipped off-site within 90 days);
- Store wastes in accordance with specified requirements for on-site storage;
- Offer the wastes only to transporters and facilities with an EPA identification number;
- Comply with applicable Department of Transportation (DOT) packaging and labeling requirements for wastes shipped off-site;
- Use a multipart "round-trip" Uniform Hazardous Waste Manifest to accompany the waste to its final destination; and
- Maintain copies of manifests for 3 years.

A full description of all the hazardous waste management requirements is available in many documents and guidebooks[4] and varies somewhat between states. Tank owners faced with storing wastes on-site should obtain pertinent hazardous waste regulations from their state agencies or EPA.

2.2.2.3 Generator Identification Numbers

Generators and transporters of hazardous wastes are required to obtain an identification number from EPA. In the case of underground storage tanks, the tank owner is the generator; although the tank cleaner may also be considered the generator and sign the manifest, some tank cleaners will not do so because of the liabilities associated with hazardous waste generation. Thus, tank owners must often obtain their own Generator ID numbers if their tank closure will produce waste gasoline or other potentially hazardous wastes.

For a single shipment of UST waste, the tank owner may obtain a Temporary Generator ID number from the State or EPA; in some states these numbers are issued by telephone while other states require the completion of an application form.

Tank owners may obtain a permanent Generator ID number by completing form 8700-12, *Notification of Hazardous Waste Activities*. This form is available from, and must be submitted to, the state hazardous waste agency or the regional EPA office. The addresses of EPA regional offices are listed in Appendix A.

2.2.2.4 Treatment, Storage, and Disposal Facilities

Hazardous waste treatment, storage, and disposal (TSD) facilities are required to have both an ID number and an operating permit (Interim or Final) from EPA. The granting of operating permits is contingent upon the facility's meeting certain specified technical and administrative requirements which ensure that the waste is handled responsibly by knowledgeable personnel. Only those firms specifically permitted by EPA or the equivalent State agency may legally engage in hazardous waste management activities.

However, due to the complex and time-consuming nature of the permitting process, facilities were first allowed to operate under less stringent Interim Status Permits until EPA requested and approved their final permit applications. Although all final permit applications were to be submitted by November 8, 1988, permit negotiations for many facilities are still continuing.

2.2.3 The Comprehensive Environmental Response, Compensation, and Liability Act (CERCLA, or "Superfund")

CERCLA, or "Superfund", is the primary federal legislation that addresses the cleanup of hazardous waste sites and chemical spills. CERCLA'S major components are a fund, raised from several sources, to finance cleanup activities; a National Priority List (NPL) of severely contaminated sites designated for cleanup under the Superfund program; and a National Contingency Plan for coordinating the government response to chemical spills. The implementation of CERCLA has been a major focus at EPA since the inception of CERCLA in 1980.

There are three aspects of CERCLA relevant to the tank owner. The first is that RCRA Subtitle I UST regulations use the CERCLA definition of "hazardous substance" to define nonpetroleum tanks subject to the UST regulations. The CERCLA hazardous substance list is much broader than the RCRA list for hazardous waste, and includes chemicals regulated by the Clean Water Act (CWA) and by other environmental laws. Any tank that contains a "hazardous substance" under CERCLA is subject to the UST regulations of RCRA, unless specifically excluded. A current list of these substances is available from EPA regional offices (Appendix A).

The second aspect is that CERCLA imposes "strict liability" on waste generators for the environmental consequences of their waste disposal. Under CERCLA, each and every party who handles a waste, including generators, transporters (if they select the disposal site), and TSD facility owners and operators can be held financially liable for cleanup costs if the disposal site is remediated under the Superfund program. When there are many users of a site, liability is partitioned among them, but under the "joint and several liability" approach of Superfund, any one user can be held responsibile for the entire cleanup cost.[5,6] The desire to reduce liability has lead some corporations with large numbers of tanks to opt for expensive but relatively safe disposal methods such as incineration, rather than landfilling, for their tank residues.

The third aspect of CERCLA important to tank owners has to do with the release of hazardous substances. Under the National Contingency Plan, any release of a CERCLA-defined hazardous substance (which excludes petroleum) in excess of stipulated "reportable quantities" must be reported to the National Response Center (NRC). This requirement pertains to nonsudden releases, such as leaks from USTs, as well as to spills and overfills. Sudden spills must be reported to the NRC within 24 hours. In addition, some large release sites may also fall under the jurisdiction of CERCLA for remediation, especially when state and federal agencies are involved; the state agency or

EPA will determine the resolution of CERCLA requirements vs. the corrective action requirements of the RCRA Subtitle I regulations (53 FR 37188).

2.3 STATE REGULATION

State involvement in UST management varies widely, but is expected to increase as more states become authorized to implement federal standards. UST and hazardous waste programs can be administered by either EPA or the state, but not by both; thus the tank owner need only be concerned with whichever agency has jurisdiction in his or her state. Some states, such as Florida and California, moved ahead of EPA and implemented their UST programs before the final regulations were promulgated, and other states, like Texas, have developed theirs in response to EPA. In any case, state programs can be no less stringent than the federal program, but may be more stringent as required to protect regional resources. Such is the case in Texas, for instance, where double-wall piping and tanks are required because of the sensitivity of the aquifer underlying some of the state.

In addition, decisions regarding key components of UST management, such as site assessment requirements, were intentionally left to state legislation. State regulations or program information can be obtained by calling the state UST offices listed in Appendix B.

2.4 LOCAL REGULATION

In some states, most notably California, UST regulation is handled almost entirely at the county level, with tank requirements varying with the sensitivity of the site. Because EPA believes that tank regulation is most effectively carried out at the state and local level, a trend towards increasing local involvement can be expected as communities respond to the federal initiative (53 FR 37096).

The state UST office may be contacted for more information about local and state jurisdiction over UST management.

2.4.1 Building Codes

Regulations governing the installation, removal, abandonment, temporary closure, and similar safety-related issues have been part of local building codes in some areas for many years. The Building Officials and Code Administration (BOCA) periodically issues National Fire Prevention Codes which are adopted by numerous states and localities nationwide. The 1990 version of these codes includes guidelines for tank closure.

Local building codes often specify when a tank may or may not be abandoned in place, and may require the acquisition of a permit from the local building inspector or codes and enforcement office.

2.4.2 Zoning Laws

In some areas, local zoning laws have been used to advance standards of construction and closure that achieve a desired level of environmental or public health protection. While this is not an especially common route of UST management, tank owners should nonetheless contact local zoning officials prior to closure, especially abandonment in place.

REFERENCES

1. Anon., Concrete basins are not tanks, rules courts, *Focus*, Hazardous Material Control Research Institute, May 1990, p.5.
2. Weston, Roy F., Inc., *Remedial Technologies for Leaking Underground Storage Tanks*, Lewis Publishers, Chelsea, MI, 1988.
3. O'Brien & Gere Inc., *Hazardous Waste Site Remediation*, Van Nostrand Reinhold, New York, 1988.
4. McCoy and Assoc. Inc., *RCRA Regulations and Keyword Index, 1989 Edition*, McCoy and Assoc. Inc., Lakewood, CO, 1989.
5. ERT Inc. and Sidney & Austin, Inc., *Superfund Handbook*, ERT, Inc., Concord, MA and Sidney & Austin, Chicago, IL, 1987.
6. U.S. Congress, *Technologies and Management Strategies for Hazardous Waste Control*, Office of Technology Assessment, Washington, D.C., 1983.

CHAPTER 3

Tank Closure: Strategies and Procedures

Janet E. Robinson

CONTENTS

0-87371-402-4/93/$0.00+$.50

Tank Closure: Strategies and Procedures

3.1. INTRODUCTION

The available means for underground tank closure are represented by a well-established set of technologies described by a number of industry standards. While innovative developments have refined some aspects of tank waste treatment and disposal, the available options for underground tank closure consist of essentially two general sequences, defined here as the following strategies: (1) abandonment-in-place; and (2) tank removal, followed by either landfilling or recycling the tank. Federal regulations allow a third, albeit temporary, option of retrofitting. Although differentiated largely by the final disposal location or treatment of the used vessel, both abandonment and removal include certain standard procedures which, when coupled with the waste management requirements of the Resource Conservation and Recovery Act (RCRA) legislation, provide a fairly straightforward sequence of operations. Variability is introduced by site conditions, presence or extent of contamination, proximity of structures, additional state or local requirements, and site-specific concerns.

This chapter describes each of the required steps and available options for the disposal of underground storage tanks (USTs). Some operations, such as tank cleaning and vapor-freeing, must be performed regardless of whether the tank is abandoned or removed. Other steps, such as filling an abandoned tank or discarding a removed one, require decisions about particular materials or technologies which must be made or approved by the tank owner prior to contracting the tank disposal.

3.2. ABANDONMENT OR REMOVAL?

As discussed in Chapter 1, the choice of whether to remove a tank from or abandon it on site resides largely with the tank owner, since federal regulation allows both, provided that the tank is cleaned. However, many state or local regulations may prohibit abandonment or restrict it to certain situations only, so the tank owner should be sure to check with state and local UST offices prior to making a final decision.

Table 5 compares the relative advantages and disadvantages of removal and abandonment. Probably the major argument against tank abandonment on site is its effect on property value and versatility; even if environmentally safe, a buried tank remains a lasting obstruction to future excavation for buildings, pipelines, utility trenches, and other routine industrial and commercial needs.

Table 5.   Removal Versus Abandonment in Place

Advantages	Disadvantages
Removal	
Removes potential source of future contamination	Excavation disturbs site and may result in a loss of business
Allows inspection and sampling of excavation	Excavation may affect structure or foundation of nearby buildings
Future property use is unaffected	May be more expensive than abandonment
Removes target of potential future regulation	
Conserves space; new tank may be installed in or near the excavation	
Abandonment	
Minimal disturbance of site or nearby buildings	Filled tank may be an impediment to future use or sale of the property
May be cheaper than removal	Tank may be subject to future regulation
Appropriate for large concrete units where demolition and disposal would be costly	Tank excavation cannot be inspected
	Sampling below the tank may require drilling through the tank bottom or access by other means
	May be prohibited by state or local ordinances

In addition, site assessment at an abandoned tank can involve drilling through the bottom of the tank or adjacent to it to obtain soil samples; site remediation, if necessary, may require the eventual removal of the tank anyway. In many localities, special permission or a permit for tank removal is required from the fire marshal. For these reasons and the others listed, the decision to abandon tanks should take into account all the current and future costs.

Abandonment is most suitable for tanks that are adjacent to or partially underneath buildings (e.g., heating oil tanks) since the excavations necessary to remove such tanks may be difficult or impossible without demolishing part of the building. But because the residues left in an abandoned tank will eventually enter the environment, it is particularly important that the tank be thoroughly and completely cleaned before filling.

3.3 STANDARD PROCEDURES AND COMMON PRACTICES

As described in Chapter 2, federal UST closure regulations rely heavily on standard industry codes of practice to determine and define appropriate tank management procedures. Although the September 23, 1988 Federal Register lists several codes that apply to tank closure, one of the most important is the American Petroleum Institute's (API) Publication 1604, *Removal and Disposal of Used Underground Petroleum Storage Tanks* (2nd ed.; 1987). This document, which provides much useful information about tank closure and safety concerns, can be obtained from the American Petroleum Institute, 1220 L Street, Washington, D.C. 20005.

Other useful industry standards are listed below. Sources for these standards are listed in Appendix C.

- API 2015: *Cleaning Petroleum Storage Tanks*
- National Fire Protection Assoc. (NFPA) 30: *Flammable and Combustible Liquids Code*
- NFPA 329: *Underground Leakage of Flammable and Combustible Liquids* (1987)
- National Institute for Occupational Safety and Health (NIOSH): *Criteria for a Recommended Standard: Working in a Confined Space*
- Occupational Safety and Health Administration: *Hazardous Waste Operations and Emergency Response; Final Rule* (29 CFR 1910.120)

In this section, standard procedures for tank removal and abandonment are described and discussed, along with tank conditioning activities common to both.

3.3.1 Tank Removal

In most cases, tank removal is the closure option chosen by tank owners, for any of the numerous reasons cited in Table 5. While technically more complex than abandonment, removal constitutes a fairly routine procedure for the many companies specializing in it, and can be performed quickly and safely if certain procedural and safety guidelines are followed.

API Publication 1604 describes the steps in a standard underground tank removal procedure; these steps are summarized below.[1]

1. Drain and flush the piping contents into the tank. Cap or remove all product piping.
2. Remove all flammable liquid or residues using explosion-proof or air-driven pumps. The use of a hand pump may be necessary to remove the last few inches of residue. A vacuum truck, if used, should be upwind from the tank, in an area that is vapor-free.
3. Excavate to the top of the tank.
4. Remove the fill pipe, gauge pipe, vapor recovery truck connection, submersible pumps, and other tank fixtures. Remove the drop tube, unless the tank will be vapor-freed with an eductor. Cap or remove all nonproduct lines except the vent line, which should remain connected until the tank is purged. Temporarily plug all other tank openings.
5. Purge the tank of all vapors (see section on "Vapor-freeing") to free the tank temporarily of flammable vapors. Test the tank to ensure that vapors are below 20% of the lower explosive limit (LEL) of the stored product.
6. After vapor-freeing, plug or cap all accessible holes. A 1/8th-inch hole should be left on the upper surface of the tank to allow equilibration of pressure differentials.

7. Excavate around the tank to uncover it for removal. Remove the tank from the hole and place it on a level surface; chock the sides with wooden blocks to prevent rolling. Plug any corrosion holes with screwed (boiler) plugs.
   (*Note:* if the tank is to be cleaned on site or if it shows signs of releasing product or sludge, place the tank on heavy-gauge plastic sheeting to avoid spreading residues on clean areas.)
8. Prior to removing the tank from the site, it should be labeled to prevent reuse. API recommends that the following text be painted on the tank in letters at least 2 inches high:
   - TANK HAS CONTAINED LEADED GASOLINE (or diesel, or whatever the specific tank contents were)
   - NOT VAPOR FREE
   - NOT SUITABLE FOR STORAGE OF FOOD OR LIQUIDS INTENDED FOR HUMAN OR ANIMAL CONSUMPTION
   - DATE OF REMOVAL: month/day/year

   Tanks that have contained leaded motor fuel or an unknown product should be additionally labeled:

   - TANK HAS CONTAINED LEADED GASOLINE
   - LEAD VAPORS MAY BE RELEASED IF HEAT IS APPLIED TO THE TANK SHELL
9. Because residues in the tank can resaturate the tank atmosphere, check internal vapor levels again before transporting the tank.
   (Monitoring tank atmospheres is especially important when tanks are moved before being cleaned.)
   Transport tanks with the 1/8th-inch vent hole on the uppermost side. Transport tanks in compliance with all local, state, and federal requirements.

Tank cleaning, not mentioned directly in the API procedure, is required by federal law. Depending on the method used, tank cleaning may take place while the tank is still in the ground, on-site after removal, or at an off-site facility. Common tank cleaning procedures are discussed in Section 3.3.3.2.

3.3.2 Abandonment in Place

Abandonment in place is a relatively straightforward means of disposal that avoids the expense and difficulty of tank removal and transportation off-site. Although obviously requiring the commitment of land to the long-term storage of the out-of-service vessel, abandonment is often the disposal strategy of choice in those cases where removal is difficult, unsafe, or otherwise unsuitable.

The API procedures included by reference in federal regulations are largely similar to those for removal; the tank is isolated, emptied, and vapor-freed as described in Steps 1 through 5, of tank removal (Section 3.3.1). The procedure for tank abandonment recommended by API then continues as follows:

6. After purging, introduce a solid, inert material through the openings in the top of the tank. If necessary, one or more holes may be cut in the tank top if

existing tank openings are inadequate. A small amount of water added to the sand or inert material as it nears the top of the tank will help reduce coning and will wash sand to the corners of the tank. A mixture of water and mud may also be used for the same purpose when "topping off" the tank.

7. After filling, plug or cap all tank openings, except for those cut to provide access to the tank. Disconnect and cap or remove the vent line.

The tank is reburied after this procedure is complete. It is important to keep a permanent record of the tank location, date of abandonment, cleaning method, sampling results, and all other pertinent data in order to respond appropriately to inquiries by a regulatory agency or prospective property buyers. The location of the tank should be noted in building plans so that future construction at the site can be designed accordingly.

3.3.2.1 Filler Material

In the recommended procedures described above, the abandoned tank is filled with a solid material to prevent eventual tank collapse and subsidence of the overlying surface. Federal regulations do not define the type of material that should be used, stipulating only that the material and the fill procedures used at closure will (1) prevent the tank from surfacing after closure, (2) support the tank structure as it deteriorates over time, and (3) completely seal the tank and associated piping from future use as a tank system (53 FR 37183).

Sand has traditionally been the filler material of choice because of its ready availability and low cost, although mixtures of sand and earth or sand and rock have also been used. The major problem with sand is the difficulty of filling all tank voids and connecting lines. Typically some water must be added during filling to spread the mixture throughout the tank and prevent coning beneath the fill hole, but the amount used must be kept to a minimum to prevent the acceleration of rusting in the steel walls.

Water has also been used as a filler material, especially in concrete tanks, although the limitations of this method as the tank ages and leaks are obvious. Unless the tank was thoroughly cleaned prior to closure, current owners of these old tanks may be faced with disposing of many thousands of gallons of what is now considered hazardous waste.

In recent years, however, two alternatives to sand have been developed that offer various advantages over the use of natural materials. Although sometimes more expensive than sand and not approved for use by all states, the ease of use or other distinctive characteristics of these alternatives serve to extend the range of situations in which abandonment is an attractive option. These two types of filler material are concrete slurry and polyurethane foam.

Concrete slurry is sold under a variety of trade names. It is a concrete product that is mixed on site and poured into the tank, where it evenly fills the tank and all connecting piping. Different slurry mixtures are available for tanks directly accessible from above or those located under buildings, where the slurry must be fed to the tank through a flexible hose.

Versions of concrete slurry are called "lean mix backfill", "controlled density fill", "flowable mortar", or "controlled low strength materials".[2] Generally, concrete slurry is a mixture of various quantities of portland cement, water, and selected aggregate materials, typically fly ash and sand.[2] Although the relative amounts of these components can be adjusted to achieve desired characteristics of strength, compactability, and pumpability, the proportion of cement is generally relatively small so that the finished product is "neither a low-strength concrete nor a soil cement, but has properties similar to both".[3]

The primary advantages of concrete slurry for underground tank closure are its ease of handling and its high density. The semiliquid nature of the slurry material allows it to fill the tank evenly and completely without the use of additional liquid, and after hardening, no residual water normally remains to promote rusting. In addition, the relatively high strength of the hardened fill prevents compaction and settling even after the confining walls of the tank rust away; however, excavation can still be accomplished with conventional equipment.[2] Initial costs may be higher than if natural materials were used, but the reduced time and labor costs associated with placing the material may largely offset the initial outlay.

Polyurethane foam is a light, inert foam material used for roofing, sheeting, and, most recently, UST filling. The material is introduced into the tank as a liquid, conforming to the shape of the UST and filling connecting pipes and voids. As the material hardens, it expands and is forced out the fill, gauge, and vent pipes.[4] According to the manufacturer, polyurethane foam (marketed under the trade name "Petro Fill"), resists both shrinking and swelling with age, is chemically inert, and is insoluble in water and in most organic solvents.[4] Since polyurethane foam is relatively lightweight, an abandoned tank filled with foam may be removed at a later date more easily than tanks filled with sand or concrete slurry.[4] However, not all tank removal contractors are familiar with the use of polyurethane foam, and some cite as a disadvantage the limited availability of Petro Fill, in comparison to the fairly ubiquitous concrete slurry.

Regardless of the type of material selected to fill an abandoned tank, recommended procedures still require the removal of the remaining product and the proper disposal of the residue. Cleaning and pumping techniques are discussed in Section 3.3.3.2.

3.3.3 Tank Conditioning

Tank conditioning refers broadly to those activities that render a used tank suitable for disposal, reuse, or recycling. Generally this includes vapor-freeing, to remove toxic and flammable vapors within the tank; internal cleaning, to remove accumulated sludge and scale; and, if necessary, additional cleaning to remove internal linings or external coatings. Although all tanks must be vapor-freed for the safety of workers during either abandonment or removal, the level of cleaning required varies with the intended "final fate" of the tank; tanks disposed of directly in a hazardous waste landfill, for instance, require less cleaning than those destined for recycling or reuse.

3.3.3.1 Vapor-Freeing

Because of the explosive and toxic nature of gasoline fumes, their removal constitutes a particularly important aspect of tank disposal. For the purposes of preventing explosion, fume concentrations must be either below the lower explosive limit of 1% hydrocarbon vapor by volume in air, the point at which the mixture is too lean to propagate flame, or above the upper limit of 10%, where it is too rich. Because of local variations in concentrations the minimum value is preferred, with a 10% safety margin; for gasoline, a 0.1% vapor level is thus considered safe for torch work. However, this level is still toxic and hence requires the use of protective equipment. Other chemicals and petroleum products have different flammable limits, which should be determined before commencing work on the tank. The vapor concentration in the tank can be checked with a combustible gas indicator (CGI) or other appropriate instruments to determine when the tank is gas-free. As an extra precaution against explosion, some tank removers prefer to reduce oxygen levels to nearly zero as well.

Currently, three methods are recommended by the API for the removal of vapors from storage tanks.[1] They are summarized as follows.

1. Force out vapors with carbon dioxide (dry ice) or nitrogen — These can be introduced as a gas at the bottom of the tank or, in the case of carbon dioxide, distributed as ice evenly on the bottom of the tank, so that the gas forces out flammable fumes as the ice vaporizes. Rendering a tank inert with dry ice is a relatively easy and commonly used method of vapor removal, provided a source of dry ice is available. The use of both dry ice and nitrogen has the advantage of simultaneously removing both flammable vapors and oxygen, two of the necessary components of combustion.

2. Ventilate with air — Using an eductor-type air mover or an air blower, tank vapors may be sucked or forced out of the tank. An eductor, operated with compressed air, is attached to the fill pipe, and draws fresh air into the tank through the vent pipe and other openings. An air blower is connected to a perforated air-diffusing pipe installed in place of the fill pipe; fresh air is blown into the tank and escapes through other tank openings. Both apparatus should be properly bonded before use, and the eductor pipe should be fitted with a 12-ft. extension to prevent the discharge of vapors near ground level. Some contractors prefer ventilating with air because they find it to be faster than dry ice, and an air compressor is often on the site for other purposes.

3. Fill with water — As the tank fills, flammable gases will be expelled through other openings. Petroleum product will float on the surface of the water and can be siphoned off separately, but the entire volume of water must be disposed off appropriately. (Note: because of the large amounts of potentially contaminated wastewater generated, the total cost of this method is often prohibitive.)

With all methods, expelled petroleum fumes, which are heavier than air, will tend to accumulate near the ground in the vicinity of the tank, potentially posing a significant safety threat in sheltered areas.

After degassing, the vapor concentration in the tank can be checked with a combustible gas indicator to determine if additional treatment is required. Because tanks with any residual product in them will tend to resaturate the internal volume over time, many contractors prefer to complete all cleaning and cutting operations immediately after vapor-freeing is complete. Some contractors will flow gaseous nitrogen through the tank during cutting to remove accumulating fumes or oxygen.

3.3.3.2 Tank Cleaning

USTs covered by federal regulations must be cleaned "by removing all liquids and accumulated sludges" [40 CFR 280.71(b)]. Tank cleaning takes place after product removal, and can be performed with the tank in the ground, on-site after tank removal, or at an off-site facility. Some localities, such as Portland, Maine, prohibit or discourage tank cleaning on-site. Off-site facilities utilizing specialized procedures such as steam cleaning or acid cleaning are sometimes cheaper than the same services performed on-site,[5] provided transportation costs are not excessive.

Two investigations have been conducted by or under the auspices of EPA to identify the types and amount of material typically found inside petroleum storage tanks.[5,6] The information obtained from these studies and from tank cleaners themselves indicates that the amount of residual material present in tanks is highly variable, and influenced by a number of factors. High tank age (over 12 years), low volatility product, low throughput, infrequent cleanings, cold weather, amount of water, and changes in contents have all been identified as characteristics that tend to increase residue quantities.[5] While 70 to 90% of the residual consists of contaminated product, the remaining 10 to 30% may contain a mixture of water, product-related residuals (gum, sediment, and tars), rust and scale, and small amounts of microorganisms, which metabolize the oil and water in the tank.[6] Many contractors report that gasoline tanks rarely contain sludge.

Water in a tank is not necessarily the sign of a leak. Water can be dissolved in the fuel on delivery, or can form as condensate from warm or moist air drawn into the tank through open fill tubes. As a result, a certain amount of water is found in the bottom of most USTs.[6]

An EPA-sponsored survey identified and described six different methods tank cleaners commonly use to remove solid or liquid residuals from USTs.[5] These are as follows:

- Tank entry and physical removal
- Low pressure water
- High pressure water
- High pressure steam
- Chemical cleaning agents
- Fuel cleaning

Some of the findings of the EPA survey are contained in the following descriptions of these methods.

3.3.3.2.1 *Tank Entry and Physical Removal (Hand-Cleaning)*

With physical entry, personnel enter the tank to scrub and scrape up internal residues. Many contractors use this as a primary cleaning method or as an initial step, to be followed by any of the methods described below. Workers inside the tank use scrapers, wire brushes, brooms, shovels, absorbent pads, and a variety of other implements. This widely used procedure has a number of advantages: (1) waste production is minimal, consisting only of actual tank residues and sometimes rinse water; (2) the tank can be inspected as it is cleaned, allowing particularly heavy areas to receive extra attention; (3) corners and crevices can be cleaned directly; and (4) no extra power equipment is required.

However, physical entry is not always possible for small tanks without cutting an opening, and personnel inside the tank are subject to risks from toxic lead or petroleum vapor. In fact, some cleaning companies and municipalities prohibit tank entry because of the danger to workers.[6]

3.3.3.2.2 *Low Pressure Water*

Low pressure water rinsing with a garden hose or equivalent is another simple method of tank cleaning. To prevent sparking from static electricity, the hose nozzle should be electrically bonded to the tank. This method produces contaminated water and may not be powerful enough to remove all residues.

3.3.3.2.3 *High Pressure Water*

This powerful technique, also known as hydroblasting, uses the mechanical action of pressurized water to remove residues and scale from internal surfaces. However, because weak or corroded parts of the tank wall may be damaged or broken by the force of the water, this method may not be the best choice for old tanks. High-pressure water washing produces quantities of contaminated water which must be pumped out and disposed of, although the actual amount is somewhat less than the amount normally produced by low-pressure water washing.[5] Some contractors add various detergents to the water to help solubilize oily deposits.

3.3.3.2.4 *High Pressure Steam Cleaning*

This may be performed either on- or off-site, and uses the scouring action of high-pressure, high-temperature steam to dislodge sludge and scale from tank walls. Soaps, degreasers, acids, or a variety of other chemicals can be added to the steam to facilitate cleaning,[7] although steam alone is frequently used. For on-site cleaning, portable steam generators, similar to those used for equipment decontamination, are used to generate steam which individuals inside the tank then apply with hand-held steam wands. An advantage to this method is that the tank can be inspected as it is cleaned, with effort focused on stubborn areas. Residues and condensed steam are then either pumped or removed from the tank by hand. This method is often used for tanks that will be reused to store other chemicals after cleaning.

Commercial facilities also exist which clean tanks as a primary business focus. In one steam cleaning operation,[8] empty tanks are placed on a specially drained steam pad and injected with 325° F steam via a hose inserted into the tank. The steam both vapor-frees and cleans the tank in one step, and is especially effective for small tanks, where other cleaning methods are difficult or impossible to implement. After the tank is cleaned and vapors are removed, the tank is initially cut by a remote-control plasma cutter; remaining sectioning can be completed with an acetylene torch. Liquids are treated at an on-site facility, and sludges are drummed and disposed of as hazardous waste. After sectioning, the tank is clean enough for sale to a scrap metal dealer.[8]

3.3.3.2.5 Cleaning Agents

Various solvents, detergents, or acids may be used alone or in combination with the above methods to solubilize particular residues. Cleaning agents include biodegradable solvents, degreasers, and emulsifiers, and acid or caustic agents such as trisodium phosphate or nitric acid, which have been used to remove lead from the scale and surfaces of leaded gasoline tanks.[5,9] Like steam cleaning, acid cleaning is often performed at specialized treatment facilities to ensure proper handling of the tank residues and waste chemicals.[5]

3.3.3.2.6 Other Methods

Fuel cleaning involves using petroleum itself as a solvent or scouring agent. A relatively lightweight fuel is used to scour and dissolve residues of heavier grades of petroleum: No. 2 oil, for instance, may be used to scour tanks containing Nos. 4, 5, or 6 oil. The fuel/sludge residues are then pumped into a truck for disposal or recycling. An alternative method is to circulate residual fuel already in the tank through a high pressure nozzle in the tank to hydraulically scour deposits from tank walls. Both the fuel and the loosened debris are continuously pumped out of the tank, with the fuel refiltered and returned to the nozzle. The final result is clean fuel in a clean tank.[5] Both methods are normally carried out with the tanks in place. Neither method is suited for tanks which have confirmed or suspected leaks in the area to be cleaned, since fuel may be driven through tank holes into the surrounding soil.

Sandblasting or other forms of abrasion cleaning may occasionally be necessary for tanks with resistant deposits or for those that have linings or coatings. By using the appropriate abrasion media, tank surfaces can be reduced to bare metal, suitable for recycling or relining. However, the spent sandblast media contains fine particles of the scoured material, and in some cases may constitute a hazardous waste if the scoured residue contained hazardous constituents or was itself a listed hazardous waste. The use and recycling of steel shot reduces the amount of used abrasive that will require disposal.

3.3.3.3 Tank Cleanliness: How Clean Is Clean?

No federal standards currently exist defining either how tank cleaning must be performed, or to what extent tanks must be cleaned, although some state and local codes have their own requirements. Contractors usually have their own standard procedures which include one or more of the methods described above, but the tank owner can specify either a method or a cleanliness standard different from those normally used. In the absence of regulatory guidance or tank owner preference, tank cleaning efforts may be based on the eventual final fate of the vessel; thus, relatively little cleaning may be required to dispose of a tank in a hazardous waste landfill, while extensive surface treatment may be necessary for a tank destined for recycling, relining, or reuse, especially if the tank has internal linings or external coatings. Many metal recyclers are unwilling to accept tanks containing chemical residues because of the liabilities associated with handling potentially hazardous waste; thus their requirements need to be met if recycling is anticipated.

Despite the lack of federal standards, the current, informal EPA policy is to consider tanks as containers which must be emptied prior to disposal; the RCRA definition of an "empty container" over 110 gallons in size is one that contains "no more than 0.3% by weight of the total capacity of the container" [40 CFR 261.7(b)(1)(iii)(B)]. As noted above, this level may not be acceptable to all tank recyclers, and the requirements of the party accepting the tank for disposal or recycling may end up being the source of functional cleaning standards.

Tank cleanliness is currently determined in a number of ways. The simplest is a visual check to see that the majority of wastes have been removed. In other cases, the standard triple rinse required by RCRA for hazardous waste tanks may be followed, with the tank deemed "clean" when the process is complete.

More rigorous methods, which may be used when documenting that a tank's decontamination is particularly important, consist of either collecting a wipe sample of the tank's internal surface after cleaning, or collecting a sample of the final rinsewater, or both.[6] Wipe sampling consists of rubbing a piece of wetted sterile gauze or filter paper over a specified amount of surface area, and then analyzing the gauze or filter paper for specified parameters. Though sometimes useful for a before-and-after assessment, disadvantages include poor reproducibility of results and the lack of standards for the type of material which may be present on the sampled surface. Analytical results from rinsate samples are commonly compared to drinking water standards, specifically the Maximum Contaminant Levels (MCLs) (40 CFR 141.11 and 141.61). Such a comparison, while seeming extreme, simply measures the amount of material leaching into rinsewater which, if supplied from a public utility, met drinking water standards to begin with.

3.3.4. The Real World: Common Problems and Variations

As with most things, the real world routinely presents situations quite different from those represented in the text of standard procedures. This is particularly true with underground tanks, which cannot be seen and may be forgotten for decades. The rule is to expect the unexpected, and to allow time in the schedule for unanticipated problems.

Some of the commonly encountered "variations" from the ideal tank removal are described below.

3.3.4.1 Contamination

Most underground tanks have at least a small amount of contaminated soil around the fill pipes or other areas, resulting from occasional overfills or spills. In many, if not most, cases involving older tanks, soil volumes ranging from a few cubic yards to many tons may need to be removed to capture the effects of periodic overfills; leaks may result in full-scale remediation efforts costing many tens of thousands of dollars. In either case, the issue is not likely to be resolved quickly because of sample turnaround time and disposal arrangements. When preparing closure arrangements, tank owners would be wise to prepare mentally and logistically for the possibility of contamination.

In a growing number of states, financial assistance is available for tank owners who incur large costs as part of required tank removals. Research into available funds should be conducted as part of the closure preparations, as much for the tank owner's peace of mind as for financial planning.

3.3.4.2 Shallow Groundwater

It sometimes happens that a tank is located, initially excavated and pumped out, but on inspection the following morning (or sooner) is found to be full of liquid again. This occurs when a tank has holes below the level of the water table; water flows into the tank as soon as the product is pumped out, with the water itself becoming contaminated with residue. The water must then be pumped out and properly disposed of, and a means found to prevent the continual refilling of the tank. One way to do this is to raise one end of the tank and pump from the lower end, lifting the tank higher as it is emptied. Regardless of the technique employed, a situation such as this results in delay and a higher-than-anticipated waste disposal cost.

3.3.4.3 Utilities

Public utilities, including overhead lines and subsurface piping or communications cables, can throw an element of the unexpected into any tank job. While a site visit usually indicates possible interference by overhead lines (assuming the tank is where it is supposed to be), the location of subsurface lines must be identified by the utilities themselves, whom contractors are required to notify before digging. In some cases, even these identifications may be found to be incorrect. Utility or process lines on private property are the responsibility of the property owner, and should be discernable on site blueprints, if they exist. In any case, the discovery, or in the worst cases, rupture of an underground utility can have serious consequences for the job budget, the schedule, and the safety of the site personnel, not to mention the reputation of the contractor.

3.3.4.4 Open Holes

When tank excavations are sampled as part of a site assessment, the delay while awaiting sample results means that an open hole is left on the tank owner's site, which presents a safety hazard and may affect retail sales. This problem may be ameliorated in a few different ways, although none is without drawbacks. Sample analysis may be expedited from the normal 30 days to a week or so by paying premium laboratory rates, 150 to 200% of the normal. Alternatively, the hole may be filled in, with the knowledge that reexcavation may be necessary if further sampling or the removal of a "hot spot" is required. Plastic sheeting may be used to separate clean fill from the original excavation to avoid contaminating otherwise acceptable fill. Scheduling tank closure during periods of anticipated low sales will minimize the effects of site excavation.

3.3.4.5 Tank Location

Tanks, especially abandoned ones, may not be where they are expected to be. "As-built" plans, or more commonly, preconstruction blueprints are often inaccurate, as may be the memories of personnel. To avoid rupturing a tank, exploratory excavation is normally done fairly gingerly; locating a tank without specific plans can therefore be time consuming. Geophysical methods, described in Chapter 4, may locate tanks more quickly than probing around with a backhoe. Imprecise location identification results in larger areas of broken asphalt than anticipated and possible interference from utilities.

Accurate tank location is particularly important when old tanks are located alongside of new, active ones, which can easily be damaged if unexpectedly encountered. A good backhoe operator with a "gentle touch" is invaluable in these situations.

3.3.4.6 Tank Size

Often tanks are not the size or material they are said to be in company records. While the difference in tank size normally does not in itself present difficulties, an important consideration is whether the discovered tank is in fact the vessel referred to in the records, or a different, unrecorded tank encountered by coincidence. How many more are there? Where *is* the tank originally sought?

3.3.5 A Note on Remediation during Removal

As noted above, many, if not most, tank removals are accompanied by the discovery of contaminated soil in the bottom of the hole, often underneath the fill pipe. In this case, standard procedure is to excavate the soil until either (1) no further sign of it remains, or (2) some logistical problem, like backhoe reach or building proximity, prevents further excavation, or (3) it becomes clear that a large amount of soil is affected.

Both tank owner and contractor should be prepared for a delay at this point. As soil is removed from the hole, the consultant, state regulator, or whoever else is conducting the site assessment will examine the soil for odors or other signs of contamination. If no further sign is found, a sample is taken and excavation stops; if contamination persists, however, excavation must continue, foot by foot. Tension often rises during this process, and unfortunately the individual evaluating the soil is often regarded as if he or she were to blame for the growing pile of contaminated soil. When and if a point is reached where the soil appears to be "clean", it is important to remember that the soil may not be clean at all, especially with relatively nonvolatile compounds; the final assessment can only be made with laboratory results.

Excavated, contaminated soil should be placed on heavy plastic in a separate pile from uncontaminated soil, and covered with plastic after excavation is complete to prevent the generation of contaminated runoff.

3.4 TANK DISPOSAL

The disposal of used USTs, once a routine matter involving either the local landfill or scrapdealer, has become increasingly problematic as the former contents of the tank come under greater regulation as hazardous or industrial waste. While some tank removers will arrange for tank disposal as part of their services, legal liability for the impacts of the disposed-of tank still resides with the tank owner, so tank owners must take an interest in the fate of their old vessels, especially if there are many of them. For tanks that will not be relined or reused, two disposal alternatives exist: recycling and landfilling.

3.4.1 Tank Disposal: Recycling

For many reasons, the recycling of old steel vessels as scrap metal is the most desirable form of tank disposal. Melting down an old tank virtually eliminates the liabilities associated with tank residues, and sale of the tank to a scrap dealer may provide a small financial return to the tank owner or contractor. In addition, recycling tanks retains for society the use of a valuable material, which may be recast into new products. Prices paid for scrap metal vary widely, depending on local scrap supply and the proximity of processing mills.[10]

The hazardous waste laws of RCRA, however, have had significant impact on the willingness of scrap dealers to accept used USTs. Because the residual sludge on the inside of an unclean or poorly cleaned UST may constitute a hazardous waste, scrap dealers may find themselves inadvertently, and illegally, handling hazardous waste if they accept such a tank on their site. In addition, the lead associated with sludge from leaded gasoline presents a serious health hazard when the tank is cut up for transportation, a risk many recyclers are reluctant to incur. For these reasons, a growing number of scrap dealers are refusing to accept used USTs, or will accept them only if the contractor, tank owner, or somebody certifies that the tank is empty and has been cleaned by approved methods.

In addition, the demands of the steelmaking process also impose stringent standards for tank conditioning. A brief examination of the factors important in metal recycling will help to elucidate both the position of the scrap dealer and, consequently, the extra needs of the tank owner who desires to recycle tanks.

The scrap metal industry consists of essentially two components: the processor, who buys, separates, and generally delivers scrap metal, and the steel mills and foundries that melt down the scrap for recasting into new products. Because of both quality controls on the finished products and stack emission limitations imposed by air pollution standards, steelmakers are sensitive to contaminating elements in their supply of purchased scrap. Some contaminants, such as copper, tin, and nickel, will negatively affect the properties of the recast steel, while others, such as lead, zinc, and aluminum will oxidize in the furnace and require removal by expensive air pollution control equipment, which in turn produces a hazardous sludge. Since steelmakers will not buy scrap that is likely to cause problems in their manufacturing process, processors must incorporate these same restrictions into their specifications for metal purchased from the public.

For the petroleum tank owner, these restrictions affect cleaning costs in two ways: linings and lead residuals must be cleaned from the inside of the tank while, in many cases, coatings must be removed from the outside. Because tetraethyl lead from leaded gasoline tends to accumulate in scale and rust deposits as well as in residual bottom sludge, processors may require that these components be removed, and the surface in contact with sludge scraped down to a bare metal surface. Although some scrap processors will clean purchased metal to varying degrees on their premises, such services will result in a lower purchase price for the tank owner. Owners contemplating recycling must thus weigh the advantages of reductions of financial liability against the added expense of reducing the tank to bare steel; this expense may be significant if, for instance, the tank is coated with asphalt. However, this factor may be of less importance now than in the future, since many older tanks are of bare steel to begin with. It is interesting to note that many of the features now installed on new tanks to reduce corrosion, such as zinc anodes and epoxy coatings, will subsequently make them unsuitable for disposal as scrap metal because of excessive conditioning costs.[10]

In addition, because of the danger involved in cutting and handling empty gasoline tanks, many scrappers require tank owners to cut up their vessels before delivery.

Despite these constraints, recycling as a disposal technology for underground vessels remains a viable and desirable option, especially for uncoated steel tanks that are free of rust and scale. Tank removers often know which scrap yards will and will not accept old USTs, and can clean the tank accordingly. Additionally, a market also exists for old aluminum tanks, but the same restrictions against coatings, linings, and surface contamination of the metal will apply.

3.4.2 Tank Disposal: Landfilling

In some cases, where the condition of the tank or distance to the scrap dealer discourages recycling, tank disposal by deposition in an industrial or hazardous waste landfill may be the preferred disposal option. Although the tank then remains a rusting source of potential contamination for many years, a well-designed landfill is built to contain or mitigate the effects of all deposited substances, many of which are more harmful than petroleum residues. And for tanks of material other than steel (e.g., fiberglass), or that have contained substances that render the steel unsuitable for recycling, landfilling is often the only option for disposal.

The requirements for tank cleaning and conditioning are rather less stringent for landfilling than for recycling. For industrial landfills (those permitted to accept industrial, but not hazardous, wastes), tanks typically must be drained of liquid and free of hazardous sludge; if sludge is present it must be shown by analysis to be nonhazardous. Because of space and safety considerations, operators often require that tanks be cut up or that the ends be cut off to facilitate compacting. A hazardous waste landfill will accept both the tank and nonflammable sludge; an absorbent filler is often then poured into the tank to immobilize the residues and prevent collapse of the structure. Disposal fees for a hazardous waste landfill are relatively high.

The financial liability of disposing of USTs in a landfill can be significant, however, if the landfill leaks and becomes involved in costly cleanup and litigation. In one mathematical model of UST landfilling costs produced by an automobile manufacturer, the financial liability of sharing in landfill investigation, cleanup, fines, lawsuits, and other facets of the site remediation was calculated to be $354 per ton.[11] While not as great a factor in the disposal of a single tank, liability becomes a prime consideration for owners of large numbers of tanks who wish to choose a consistent disposal route.

Sanitary landfills, intended primarily for household trash, range from unlined holes to, in some newer facilities, highly engineered systems with liners and leachate collection systems. Although sanitary landfills were commonly used as disposal sites for old underground tanks before the advent of the current waste management laws, landfill operators are now becoming increasingly wary of dirty USTs or anything else that causes environmental problems at their facility. As a result, the availability of the sanitary landfill as a disposal option is diminishing, although some landfill owners will still occasionally accept clean, cut-up vessels. Because of the lower level of environmental protection offered by these facilities, however, they are not recommended as an appropriate disposal location for used tanks.

3.5 WASTE DISPOSAL

The closure of USTs usually produces four general kinds of waste products, in addition to the tank itself:

- Excess product
- Sludge, rust chips, and other tank residues
- Trash (discarded plastic sheeting, protective clothing, etc.)
- Contaminated soil

Depending on the contents of the tank, the disposal of these waste streams may be problematic, since some of these waste streams edge up to the definition of "hazardous waste" under federal and state laws. Understanding fully what is and what is not a hazardous waste is important to ensure that disposal routes do not run afoul of applicable laws.

3.5.1 Definition of "Hazardous Waste"

Under federal definition, a discarded material is regarded as a hazardous waste for one or more of three reasons:

1. It is specifically listed as a hazardous waste by the EPA (i.e., it is a "listed hazardous waste")
2. It meets or exceeds EPA-defined characteristics of corrosivity, reactivity, toxicity, or ignitability (i.e., it is a "characteristic hazardous waste")
3. It is mixed with a listed hazardous waste.

With the exception of tank bottoms from leaded fuel storage tanks in the petroleum refining (not distribution or sale) industry, petroleum or petroleum sludge and residues are not "listed" hazardous wastes under federal definitions.

However, some states, such as Massachusetts, do classify used oil as a hazardous waste, requiring petroleum tank residues to be handled and disposed of according to hazardous waste regulations.

To meet the characteristic of corrosivity, the waste must be either highly acidic or caustic, with a pH less than 2 or greater than 12.5. Petroleum products generally do not meet these characteristics, although spent acids used to clean the insides of tanks may. Cleaning acids are generally used at specialized treatment facilities, however, so the treatment facility is responsible for disposing of the acidic or caustic waste, not the tank owner.

Reactive wastes are those which are generally unstable; they react violently with water, explode under standard conditions or when exposed to heat, release toxic fumes at extreme pHs, or exhibit other reactive properties. Most chemicals stored in tanks do not fall into this category.

Toxic wastes are those which have levels of organics and metals which are either present in liquid waste or capable of being leached out of a solid waste, when the solid waste is subject to a standardized "Toxicity Characteristic Leaching Procedure" (TCLP) established by the EPA. The TCLP, along with standards for benzene, toluene, and other organics, were promulgated in 1989 as a revision of an earlier toxicity standard which included only eight metals and used a different leaching procedure, the "EP (Extraction Procedure) Toxicity Test." Under the new standards and TCLP, nearly all petroleum sludges and residues would be classified as hazardous wastes.

However, because of the uncertain impact of this designation on the federal UST program, EPA has ruled, at least for the time being, that gasoline or petroleum-contaminated groundwater, soils, or other materials associated with corrective action for USTs are not hazardous wastes, even when the standards for metals or organics are exceeded. This deferral may change in the future.

The fourth group of characteristic hazardous wastes, ignitable wastes, are generally those wastes which have a flashpoint over 60°F (the "flashpoint" is the temperature at which a material will ignite when exposed to a flame or spark), or which, under standard conditions, can cause a vigorous and persistent fire as the result of friction, absorption of water, or spontaneous chemical changes. Under this standard, many liquid or slurry wastes from petroleum USTs may be classified as hazardous if significant amounts of volatile components are still present. Sludge, however, may not have enough volatiles to meet the ignitability criteria, and so may not be a hazardous waste; for this reason sludge and waste product should be drummed separately. The ignitability test does not apply to solid wastes such as soils.

Thus, petroleum-derived wastes have the potential to be hazardous wastes because of their ignitability or because they are categorically included as a "listed" hazardous waste under state regulations. Other chemicals would be categorized in accordance with their specific characteristics.

Note: Nothing is a hazardous waste until it becomes a waste by being "discarded", i.e., is "abandoned, recycled, or inherently waste-like" [40 CFR 261.2(a)(2)]. Thus, free product in a tank may be pumped out and used; it need not be discarded simply because the tank is being closed.

3.5.2 Petroleum-Contaminated Soils

Petroleum-contaminated soils, which fall into a gray area under federal regulation, are covered primarily under state laws, which vary considerably from state to state. State positions generally fall into one of three categories:[12,13]

1. Petroleum-contaminated soil is not a hazardous waste, because petroleum is not a listed hazardous waste
2. Petroleum-contaminated soil is a hazardous waste, because petroleum contains RCRA-defined hazardous constituents such as benzene and toluene
3. Petroleum-contaminated soils are only hazardous waste if they meet hazardous waste characteristics

In one study conducted in 1988, 25 of 28 states surveyed regarded petroleum-contaminated soil as described in Category 3, above.[14] The hazardous waste characteristic of interest, however, was toxicity as determined by the former EP Toxicity test, in force until the advent of the TCLP in 1989. However, not one of thousands of EP Toxicity tests conducted by various states on petroleum-contaminated soil exhibited the characteristic of EP Toxicity, a result which is not surprising in light of the high adsorption coefficient of lead to clay and organics in natural soils.[15] Although the TCLP is known to leach metals more aggressively than the EP Toxicity procedure, the exemption of UST residues from the TCLP leaves the criteria of hazardous waste characteristics, and the position of states that rely upon it, in question.

In addition, many states have informal or formal numerical standards by which they evaluate the success of cleanup programs. In some states, if the state inspector cannot see or smell petroleum product in the soil, the site is considered clean in the eyes of the state agency. Like the characterization of petroleum-contaminated soils, these cleanup levels vary between states and even within the same state over time, as UST programs evolve. For information in both areas, the best idea is to contact state officials directly.

Strategies for disposing of petroleum-contaminated soils also vary widely between states.[12] In 1988, the most common repositories were landfills, both hazardous and nonhazardous; landfarming (mixing contaminated soil in with native soil for natural biodegradation) was also allowed by about two thirds of the states at that time.[12] Other options nationwide included incineration, use as cover or road fill at landfills, landspreading, and treatment methods such as bioremediation or soil venting.[12] Some states will allow the tank owner to spread the soil at the tank site or other available property if the site is suitably rural and meets other standards. Local landfills may take limited amounts of soil. Minnesota and several New England states allow soil treatment at asphalt batching plants.

A note on disposal: when petroleum-contaminated soil is transported to another state for disposal, it is the regulations of the state in which the disposal facility is located that determine how the soil will be disposed of, rather than those at the original tank location. Thus, soil disposed of in a state that considers petroleum contaminated soils to be hazardous must be disposed of as hazardous, even if the soil originated in a state that does not classify petroleum-contaminated soil as hazardous.

State regulators are often able to provide information about permissible treatment and disposal options for petroleum-contaminated soil. In addition, tank removal firms also may provide information, since they are often contracted to dispose of contaminated soil that they remove.

3.5.3 Waste Disposal Responsibilities

Under CERCLA ("Superfund") the ultimate responsibility and legal liability for waste disposal resides with the tank owner. Thus, it behooves the tank owner to take an interest in the eventual treatment or disposal of the tank residues.

In many cases, the tank removal firm will be able to arrange for disposal of the tank residues for a flat per-gallon charge. In order to ensure that the material is treated or disposed of appropriately, the tank owner should investigate the TSD facility used to ensure that it has the necessary permits to handle the waste. Regardless of who is paid to arrange for waste disposal, *the tank owner can still be held liable* if the waste is disposed of illegally, so it is good practice to verify that all parties handling the waste are authorized to do so by the state regulatory agency or EPA.

In cases where the tank remover does not offer sludge disposal services, arranging for the transport and disposal of the waste material will fall to the tank owner directly. Correctly identifying the waste and arranging for disposal is often time consuming, but can be made easier by a basic understanding of the issues involved.

Depending on its characteristics, petroleum sludge and residues must be disposed of either as a hazardous waste or as an industrial, or "special", waste. Industrial waste is regulated primarily at the state or local level, and is generally much cheaper to dispose of than hazardous waste. RCRA regulations allow the waste generator (the tank owner, in this case) to determine whether the waste is hazardous in one of two ways: (1) by sampling the sludge, or (2) by using knowledge of the waste to make the determination without sampling. In the latter case, caution is essential when a waste is declared nonhazardous, since the penalties for improper disposal of a hazardous waste are substantial.

Sampling requirements vary with the disposal facility. Industrial waste facilities (i.e., those handling industrial, but not hazardous, wastes), usually require extensive analyses, but in return charge relatively low fees for treatment and disposal, whereas hazardous waste facilities require less comprehensive sampling, but charge relatively high fees for transport and disposal. Some industrial waste facilities require no sampling at all for petroleum-contaminated soil, only a letter certifying the nature and source of the contaminated material. For a single barrel of sludge or waste petroleum product, the requirements for sampling can make the disposal costs of either waste types similar.

As the quantity of waste increases, however, the potential savings afforded by industrial waste disposal make more attractive a sampling program to determine whether a waste is hazardous or not.

To dispose of a hazardous waste, a tank owner must comply with the following requirements.

1. Obtain a permanent or temporary Generator ID number from the EPA or the state hazardous waste regulatory agency (see Appendices 1 and 2 for addresses and phone numbers).
2. Store the waste on-site in compliance with state or federal requirements.
3. Transport the waste via an authorized hazardous waste transporter who has a Transporter ID number from the EPA or state regulatory agency.
4. Treat or dispose of the waste only at a facility in possession of a valid Part B permit from EPA or the state regulatory agency authorizing it to treat, store, or dispose of hazardous waste (ask for a copy of the permit).
5. Fill out a hazardous waste manifest when transporting the waste. The manifest will be signed by the generator, the transporter, and the TSD facility on receipt of the waste, with the original returned to the generator. This copy must be kept on file by the generator for at least 3 years.
6. Remove the waste from the site within a specified number of days from the time it was pumped from the tank, unless the tank owner has a licensed hazardous waste storage facility on-site. For "large quantity generators" of hazardous waste (those producing more than 1000 kg/month), the waste must be moved within 90 days. Hazardous waste generators producing between 100 and 1000 kg/month may store waste up to 180 days, provided the quantity stored never exceeds 6000 lbs. and other specific requirements are met (40 CFR 262.34).

The state or regional EPA office will assist new generators with proper procedures, and should be consulted to obtain accurate information, since requirements differ between states and are periodically revised. Some states publish lists of licensed transporters or in-state and regional disposal facilities.

3.5.4 Waste Disposal and Tank Owner Liability

As mentioned in Section 2.2.3, the "strict liability" clause of the CERCLA ("Superfund") law means that a waste generator is eternally responsible for any environmental impacts of his or her waste, regardless of the fact that others may have been paid to transport or dispose of it. In addition, CERCLA also imposes "joint and several" liability on users of disposal sites, which means that all users share in the liability for site cleanup, and that any one user can be held responsible for the entire cleanup cost, if others cannot pay. Thus, in order to reduce the liability associated with waste disposal, it behooves the tank owner to acquire the best possible treatment available for tank sludges and residues. For owners of large numbers of tanks, this is especially true.

Ultimately, some portion of most waste streams end up in a landfill. Modern landfills are highly engineered systems with multiple liners, leachate collection systems, and groundwater monitoring wells; through both design and operation, they are intended to minimize or eliminate the generation of leachate, and to prevent the passage of that leachate into the environment. However, since most chemical wastes do not decay or degrade in the landfill environment, landfills essentially represent what is a form of long-term storage for wastes. Because of the proximity of the waste to the environment (i.e., the underlying soils) and the indefinite time period (essentially forever) in which it will be stored, landfills are relatively vulnerable to problems associated with the release of hazardous constituents to the environment. In addition, the costs arising from the cleanup of leaking landfills can be immense, if groundwater or other natural resources are affected. For this reason, landfills are most appropriate for wastes or residues that can be treated by no other means, wastes for which the lack of other disposal options necessitates the risk associated with landfill use.

Many petroleum hydrocarbon products, however, can be treated in a number of ways, with incineration, co-combustion for energy recovery, and recycling being the primary methods. These methods constitute treatment, rather than disposal; they reduce the final volume of material to a relatively small fraction of the original quantity. The ash or untreatable residues can then be landfilled, either directly or after additional treatment. Because the volume and amount of hazardous constituents in the landfilled residue is greatly reduced relative to the original waste material, the liabilities associated with disposal are reduced as well.

Regardless of the treatment or disposal method selected, it is worth investigating the process and operational status of the TSD handling the waste material. Large corporations periodically audit the TSD facilities they use to make sure that their waste is being handled properly. While TSD audits may be beyond the means of the average tank owner, the following kinds of information, obtained through phone calls or a letter request, can help ensure that the TSD is properly managed.[16]

- **Treatment Process** — What kind of process does the facility use? Are any constituents being released by the unit? For treatment facilities, what are the wastes produced, and how are they managed? Are wastes stored on-site prior to treatment? For how long? Are stored wastes protected from weather or damage? If storage is for over 180 days, does the facility have a storage permit?
- **Landfill Design** — How are the landfill liners designed? State-of-the-art landfills consist of at least two liners, with a leachate collection system both above and between the liners. Older landfills may only have a single liner. Liners can be either of compacted clay or a variety of synthetic materials, but the more liners in the foundation and cover of the landfill, the more leak-proof and environmentally safe the unit will be.
- **Permits and Compliance** — Does the facility have a Part B operating permit from the EPA or state regulatory agency? (Ask the facility to send you a copy.) Is it valid for your kind of waste? Are there any citizen or other legal suits pending against the facility? Have any Notice of Violations (NOV) been

issued against the facility by the regulatory agency? (Contact the enforcement division of the State or EPA.) (For landfills, ask whether groundwater monitoring reports have indicated releases from the site.) For incinerators, determine whether emission standards have been met.

- **Site** — What is the geology underlying the unit? High water tables, close proximity to drinking water sources, highly fractured rock formations beneath the facility, and other similar characteristics may increase the chances for serious environmental problems if wastes leak out of the unit.
- **Insurance** — Enquire about the amount and type of insurance coverage maintained by the facility, especially those under interim status operating permits. The facility should be able to finance remedial actions in the event of a release without either the facility or EPA seeking financial recourse against the users of the facility.

Because obtaining this information can take some time, questioning should be started in advance of the closure effort. Evaluating a TSD in this way does not ensure that the facility will not be involved in costly cleanup programs in the future, but will reveal the obviously unsuitable companies. Once again, the hazardous waste generator is responsible for damages resulting from the waste regardless of the credentials or cost of companies hired to manage it; the time initially spent to investigate handling and disposal of the waste may save thousands later.

REFERENCES

1. American Petroleum Institute, Removal and Disposal of Used Underground Storage Tanks, Pub. No. 1604, Washington, D.C., 1987.
2. National Ready Mixed Concrete Association, Flowable Fill (promotional bulletin), Silver Spring, MD, 1985.
3. U.S. Department for Housing and Urban Development, Specification for Lean Mix Backfill, Washington, D.C., 1984.
4. Petro Fill, Inc., promotional literature, 1989.
5. PEI Assoc. Inc., Background Report on UST Closure: Current Practices, Draft Internal Report to U.S. EPA, Office of Underground Storage Tanks, October, 1988.
6. Lyman, W.A. and Tafuri, A., Evaluation of UST cleaning methods and characterization of residuals at closure, reviewed in *Hazardous Waste Consultant*, 10, 1–13, 1990.
7. Peterson, J.M., Safe procedures for storage tank investigation and removal, in *Proc. 6th National Conf. Management of Uncontrolled Hazardous Waste Sites*, Hazardous Materials Control Research Institute, 1985, 198.
8. Perguidi, V., Massachusetts Tank Disposal, Chicopee, MA, personal communication, 1990.
9. Hurst, T., Seaboard Chemical Corp., Jamestown, NC, personal communication, 1987.
10. Cutler, H., Institute of Scrap Recycling Industries, Washington, D.C., personal communication, 1989.
11. Anon., GE develops method for determining true costs of waste management—including future liability, *Hazardous Waste Consultant*, 6(2) 1-1, 1988.
12. Kostecki, P.T., et al., Regulatory policies for petroleum-contaminated soils: how states have traditionally dealt with the problem *in Soils Contaminated by Petroleum—Environmental and Public Health Effects*, Calabrese, E.J. and P.T. Kostecki, Eds., John Wiley & Sons, New York, 1988, 415.

13. Simpson, L.C., States set soil analysis regs. *Poll. Eng.*, 22 (8),93, 1990.
14. Bell, C.E. et al., State of research and regulatory approach of state agencies for cleanup of petroleum-contaminated soils. *in Petroleum Contaminated Soils,* Vol. 2.,73, Calabrese, E.J. and P.T. Kostecki, Eds., Lewis Publishers, Chelsea, MI, 1989.
15. 53 FR 37189; September 23, 1988.
16. Carney, M., Choosing a hazardous waste disposer. *Poll. Eng.,* 21(4), p.82, 1989.

CHAPTER 4

Site Assessment

Janet E. Robinson

CONTENTS

0-87371-402-4/93/$0.00+$.50
© 1993 by Lewis Publishers

Site Assessment

4.1 INTRODUCTION

Before permanent closure or a change-in-service is completed, owners and operators must measure for the presence of a release where contamination is most likely to be present at the UST site.

40 CFR 280.72(a)

The federal requirement for site assessment at tank closure represents a formal acknowledgment of the high potential for some degree of soil contamination around many, if not most, old underground storage tanks (USTs). While the amount of contamination is often small, usually the result of occasional overfills, a scientific assessment of the tank area or excavation is necessary to ensure that contaminated soil can be identified and remediated so that it does not remain a buried source of groundwater degradation long after the tank is gone.

From the EPA perspective, the object of the site assessment is "to measure for the presence of released regulated substances and provide a preliminary indication of the need for and scope of further corrective action activities. It is not intended to define the full extent or location of soils contaminated by a release" (53 FR 37171). This chapter, therefore, will focus on the design and execution of site investigations that meet the objective of the federal law: specifically, visual observations, and direct and indirect sampling techniques that may be used to ascertain whether a release of hazardous substance has occurred. In the event that a release is detected, a more comprehensive sampling scheme, typically involving soil borings and monitoring wells, is necessary to define the volume and areal extent of the contamination and hence to determine an appropriate remedial action.

According to federal regulations, the "measurement methodology" utilized during the site assessment must take into consideration factors such as the nature of the stored substance, the type of backfill used around the tank, and the depth to groundwater. Any other factors must also be considered that may be appropriate for identifying the presence and source of contamination from the UST system. Thus, soil gas sampling may be appropriate for volatile components, while direct sampling may be necessary for heavier hydrocarbons or inorganic constituents (53 FR 37184).

Currently, the sampling requirements of state implementing agencies vary widely, and are likely to change as state programs develop. In some states, a site assessment can be as little as a check for odors and visual contamination; a tank excavation exhibiting neither of these characteristics is considered "clean" for state purposes. Other states specify the number, type, location, and analyses of samples to be collected.[1] As with other phases of the closure process, the firm conducting the site assessment must be totally familiar with the state requirements before undertaking sampling.

The preliminary investigation techniques discussed here, although relatively straightforward, nevertheless require some of the same expertise in soil science, sampling techniques, and data analysis as do larger hazardous waste investigations. The activities involved in a UST-site assessment can be divided into the following general stages:

1. Review of records
2. Visual inspection during excavation
3. Sample collection and analysis (optional in some states)
4. Data interpretation and recommendations

These stages, along with the tasks that comprise them, are discussed in the following sections.

An important note: there is one instance in which a site assessment is not required — when the tank being closed has in operation at the time of closure an approved external release detection system that has not indicated the release of product from the tank [40 CFR 280.72(a)].

4.2 RECORDS REVIEW

As part of the site assessment, any records, official or unofficial, that exist on the use, condition, or contents of the tank should be assembled and reviewed.[2,3] These records might include engineering drawings (blueprints or as-builts), inventory records, maintenance schedules, repair receipts, spill reports, product receipts, records of previous site ownership or use, or anything else that would shed light on the historical use of and conditions around the tank. Records or personal recollections of spills — from any source — that occurred at or near the tank are particularly important. Also significant is whether the tank bed is located in native soil or in fill material transported from another site; unclean fill may have its own complement of petroleum residuals that may be mistaken for a product release. Information of this sort is essential to select appropriate analytical protocols and interpret the resulting data correctly. This information should be assembled by the tank owner and made available to the contractor conducting the site investigation prior to the initiation of closure.

Additionally, local information such as site topography, soil types, subsurface geology and hydrogeology, annual precipitation rates, location of drinking water supplies, and similar data will allow an initial evaluation of possible contaminant transport routes and public health impacts of a contaminant release in the area. This, in turn, may also influence the choice of analytical protocols.[3]

4.3 VISUAL INSPECTION

Much information about the condition, and hence the contamination potential of tanks and piping can be obtained during the closure procedure itself, prior to the actual collection of samples. Signs to look for, either before or during excavation, include the following:

- Stressed vegetation
- Fumes around the area or in below-grade locations such as basements, utility conduits, or sewers
- Corroded areas on tanks and piping, especially where piping or other metal attachments contact the tank
- Loose or improper connections or other defects
- Holes or cracks in the tank surface, especially directly below the fill pipe, where the dipstick would routinely contact the tank bottom
- Stained or discolored soil, with or without an odor, particularly around and below the fill pipe
- Oil sheen on water in the excavation
- Water in the excavation

If free product or an oil sheen is present, an attempt should be made to identify its source within the excavation; in some cases, the excavation of one tank may collect product actually leaking from an adjoining tank, or even leaching from soil contaminated by an old spill or a previously removed vessel. Observed free product different from that stored in the tank may dictate a change in the analytical protocol and the site investigation in general.

Some old tanks coated with tar or asphalt will produce a sheen on soils or water in contact with them for long periods. This potential source of petroleum hydrocarbons should be taken into consideration during sample collection, analysis, and data interpretation (see Section 4.5).

During excavation, it is also useful to observe the soil profile for signs of shallow groundwater. The presence of soil mottling, reddish iron oxide stains, graying, clay skins, and other properties of soil can be useful to the site geologist.[3] Unless the depth to groundwater is known, excavating or collecting soil to 5 ft below the tank bottom is sometimes useful to determine the presence of shallow groundwater or seasonal high water near the tank.[3]

4.4. SAMPLE COLLECTION AND ANALYSIS

The most common form of data collection at the time of tank closure is the collection of soil samples directly from the tank excavation. This task is preceded by decisions about sample type, location, collection methodology, and analytical parameters, and followed by the often complex task of data interpretation. These are each discussed separately below.

4.4.1 Sample Types and Locations

The primary objective of soil sampling is to determine whether a release has occurred and, secondarily, to determine the location within the sampled area of the contaminated soil. To some extent, the choice of sampling strategy depends on the importance of the latter objective, which in turn is influenced by the evidence or suspicion that a release has in fact occurred.

As noted earlier, many states have their own formal or informal requirements for soil sampling protocol,[1] or field screening procedures which eliminates some of the decision-making normally associated with site assessment. However, the following sections may be used to develop a soil sampling program in the event that no state guidelines exist.

Soil samples can be collected either as discrete (grab) samples of individual locations within the tank excavation, or as one composite sample consisting of soil taken from various representative locations within the excavation. Both strategies are acceptable, depending on the type of information desired, and understanding the difference between them can save considerably in either initial or subsequent analytical costs.

4.4.1.1 Composite vs. Discrete Sampling

Composite sampling is performed by collecting equal portions of soil from several locations and mixing them thoroughly to form one representative sample. The soil may be homogenized by mixing with stainless steel spoons in a large stainless steel mixing pan or bowl, by rolling the soil on a large square piece of plastic or cloth, or by other innovative methods.[4] As an alternative to submitting the entire volume of collected soil, the homogenized material may be divided into four equal portions, with equivalent amounts from each portion then used to fill the sample container.

Compositing is an effective way to obtain an estimate of the mean concentration of a soil constituent at a minimum analytical cost. Compositing is especially useful in heterogenous soils and, provided that the sample is thoroughly mixed, yields a reasonable level of precision without a large number of samples. However, the danger exists of diluting the concentration of a soil from a single "hot spot" to a level below regulatory concern or, in some cases, to below the detection limit of the analytical method. Thus, compositing is best for determining the simple presence or absence of a constituent, rather than for quantifying its concentration or identifying its location within the sampled area.[4,5] A disadvantage of composite sampling is that the extensive mixing and sample handling required may allow some petroleum constituents, such as benzene or toluene, to volatilize off the soil matrix, thus reducing the accuracy of the resulting data. This is of particular concern when sampling for gasoline contamination. Nonetheless, when used judiciously, sample compositing can be a useful way to reduce analytical costs, particularly when seeking contaminants that have a small likelihood of being present.[6]

If significant contamination is found in the composite sample, additional sampling may then be required to identify the location of the source within the excavation.

Grab sampling, on the other hand, involves collecting and analyzing separate soil samples from different locations. The total laboratory costs of this technique are higher since the number of samples analyzed is usually larger than with composite sampling, and the sampling process will be slower because of the need to decontaminate equipment between each sample collection. The primary benefit of this method is that the exact quantity and location of chemicals in a sampled "hot spot" is known exactly, which in turn helps focus additional sampling or soil excavation. In addition, the disturbance of sample soil is minimized, so less loss of volatile components will occur during sample collection.

One compromise approach is to collect a composite sample from locations where no contamination is suspected, and supplement these with grab samples from areas identified as potentially contaminated.[5] Potentially contaminated areas might be those showing visual evidence of contamination or those locations frequently affected by releases, such as beneath pipe fittings or fill pipes.

In summary, composite sampling is useful when no sign of contamination exists and sampling is conducted primarily to determine whether a release has occurred at all, with more detailed investigative efforts, if necessary, left to a subsequent phase of work. Grab sampling, on the other hand, provides information useful for an initial evaluation of the source and degree of contamination and helps to focus the scope of subsequent work.

4.4.1.2 Sampling Location

The actual sampling locations are chosen to be representative of the tank excavation as well as to evaluate any visual anomalies present in the tank area. Some states have regulations about acceptable sampling practices; these, of course, should be incorporated into the sampling scheme. In general, samples are usually collected at the site of suspected releases (stained or oily soil) and then at other representative areas, for instance, at the bottom center and two ends of the excavation. California UST guidance recommends collecting samples 1 to 2 ft below the bottom of the excavation at suspected worst-case locations.[3] Composite samples may include soil from a larger number of areas than grab samples, for instance, from the excavation bottom and all four sides.

In order to reduce the effects of cross contamination, sample collection should proceed from the areas of least to areas of greatest potential contamination. As described in more detail under "Equipment Decontamination" (Section 4.7), sampling equipment should be decontaminated between each sample collection.

For tanks that are abandoned in place, samples should be collected from beneath the center of the tank or its lowest point, if it can be determined.

Alternatively, if the top of the tank can be exposed, one sample may be collected from where the tank and piping meet, and a second from the opposite end of the tank. Samples may be obtained by cutting through the floor of the tank, or by using mechanical slant borings.[3]

In addition, samples should also be collected from piping runs, since loose or corroded piping is frequently the source of product releases. Some states have specific requirements; California, for example, requires that one soil sample be collected for every 20 ft of piping trench.[3]

In some cases, water may be encountered in the tank excavation, especially after periods of high precipitation. This water must be pumped from the excavation prior to sampling, and either containerized or, if the regulatory agency approves, released. In areas of shallow groundwater, the water may continually return to the excavation, usually filling it to a consistent level; in these cases the excavation may more nearly resemble a small pond than a former tank location. Sampling groundwater appearing in the tank excavation may thus be used to evaluate generally the effect of contamination on groundwater; however, groundwater samples cannot serve as a replacement for soil sampling in this instance if soil is inaccessible.

Finally, a certain amount of contamination is commonly found around most tanks, the result of periodic overfilling over the years; this soil is normally excavated at the time of tank closure and taken off-site for treatment or disposal. In this case, it is useful to have the excavator remove all obviously contaminated soil before collecting samples for laboratory analysis. This practice has two advantages: (1) it takes advantage of the sensitivity of laboratory instrumentation by using it to detect contamination that on-site noses or screening equipment cannot, and (2) it reduces the likelihood of sending to the lab a highly contaminated sample which must be diluted before analysis, a practice that results in a high final detection limit and thus less useful data. There is little point in sampling soil that obviously contains large quantities of a known product; the analytical money is better spent analyzing apparently "clean" soils that remain after all the obvious contamination has been removed.

4.4.2 Selection of Analytical Methods

The factors involved in selecting soil sampling parameters and in analyzing the data obtained therefrom are extensive and complex, because of the diverse factors involved as well as the evolving nature of analytical procedures and scientific understanding. This section illuminates key aspects of soil sampling analysis, primarily as applied to petroleum contamination.

Understanding soil sampling parameters for petroleum products requires an initial understanding of the characteristics of petroleum in general. A brief discussion of these will help elucidate some of the factors affecting the selection of methodology.

4.4.2.1 Petroleum Products

Petroleum consists of various kinds of hydrocarbons, which are compounds containing chains of carbon atoms surrounded by hydrogen atoms. The carbon structure may be arranged in straight chains (aliphatics), branched chains (including paraffins and olefins), or circular forms (aromatics). Generally speaking, the less volatile "heavy ends" of petroleum consist of the larger, more complex compounds.

The term "alkane" refers to straight-chain hydrocarbons that are connected by single bonds; examples include ethane, propane, and octane. The n-alkane, n-alkylaromatic, and aromatic compounds of the C_{10} to C_{22} range (e.g., those containing between 10 and 22 carbon atoms per molecule) are the least toxic and most readily biodegradable.

Petroleum products are divided into four broad categories: gasoline; middle distillates; heavier fuel oils and lube oils; and asphalts and tars.[2] Despite this apparently neat categorization, the products themselves represent a continuum of properties and constituents, with products in one category containing some of the same constituents as products in other categories.

Gasoline consists primarily of aromatics, cycloparaffins, and branched paraffins; a total of 276 different compounds have been identified, exclusive of additives.[7] Significant components of gasoline include the aromatic compounds benzene, toluene, ethylbenzene, and xylene (BTEX); these are commonly selected as indicators of gasoline contamination because of their relatively high solubility, volatility, and toxicity. These four compounds account for approximately 15% of unleaded gasoline, with benzene alone constituting around 1 to 3%.[8] Under laboratory conditions, benzene is the most water-soluble gasoline hydrocarbon, being approximately three times as soluble as toluene, and ten times as soluble as xylenes and ethylbenzene.[7]

Gasoline also contains a number of additives to enhance the octane or performance of the gasoline or to aid in product identification. Some of the more common additives include methyl-tertiary-butyl-ether (MTBE), ethylene dibromide (EDB, also called 1,2-dibromoethane), and tetraethyl (organic) lead.

MTBE is added to gasoline to reduce carbon monoxide emissions and boost the octane.[9] Depending on the blend, MTBE may comprise up to 10% by volume of gasoline.[2,9,10] As a component of gasoline, MTBE is relatively toxic, highly soluble, and very mobile; thus it is often transported in groundwater much faster than other contaminants.[10]

Tetraethyl lead is added to gasoline as an antiknock compound. EDB is present along with lead in some leaded gasoline. Both have been used as indicators of releases from leaded fuel tanks.

Middle distillates are more dense, less volatile, less mobile, and less water soluble than gasoline, and have a much lower percentage of BTEX, although they are still present in measurable quantities. Examples of middle distillates include diesel fuel, kerosene, jet fuels, and lighter fuel oils. Diesel consists largely of straight-chain hydrocarbons (alkanes and alkenes) from C_{10} to C_{23}; C_{16} and C_{17} predominate. Diesel fuel may contain aromatics as well.[2,3]

Heavier fuel oils and lubricating oils are more viscous than middle distillates, relatively insoluble in water, and hence relatively immobile in the subsurface. However, they share many of the other characteristics of middle distillates, including the presence of small amounts of lighter hydrocarbons.[2]

Asphalts and tars are familiar compounds characterized by their solid form at room temperature and nearly complete insolubility in water. Asphalt was frequently used as a corrosion-preventive coating on the outside of underground tanks.

4.4.2.2. Indicator Compounds

The standard approach to identifying petroleum releases, especially those involving gasoline, is to identify and quantify compounds representative of the stored product. As mentioned in the previous section, BTEX, EDB, and tetraethyl lead are frequently used as indicators of both unleaded and leaded gasoline. In addition, MTBE is under increased consideration.

The use of BTEX as an indicator has several advantages. As relatively mobile compounds, BTEX will spread away from the source in advance of heavier components, thus allowing early detection of releases. In addition, BTEX are readily detectable in both oil and water by standard analytical techniques, and the volatility of these compounds makes them amenable to detection in soil gas, a useful characteristic for both headspace screening of soil samples and soil gas surveillance of the site.

However, BTEX analyses are not appropriate for all gasoline spills. The high volatility of BTEX allows a certain amount to evaporate into soil pores or the atmosphere and, as relatively simple organic compounds, they biodegrade in the subsurface sooner than more complex molecules. Its solubility, while an asset for detecting groundwater contamination, allows BTEX to migrate away from the site of the spill, affecting the quantities present in point-of-release soil samples. For these reasons, BTEX is most appropriate for recent spills or releases, where the BTEX has not yet had time to biodegrade or be transported away.

EDB is used in trace amounts in leaded gasoline as a lead scavenger and thus can serve as an indicator of leaded gasoline contamination. Using EDB as an indicator should be used with caution in rural areas, however, since EDB is also widely used as a soil fumigant for pest control[11] and may be present in low concentrations at the site.[2,3] In urban areas, EDB can be a useful sign of leaded gasoline release.

Analyzing for tetraethyl lead is also a useful way to detect releases of leaded gasoline, or to distinguish leaded gasoline from petroleum products present in the subsurface. However, since some laboratories analyze only for total lead and cannot separate organic, tetraethyl lead in gasoline from inorganic lead naturally occurring in soils, a background sample of clean soil from or near the site may be required to help distinguish between the two. Given the high toxicity of organic lead, however, such extra efforts may be warranted, especially near schools or residential areas.[3]

Although not widely used yet, MTBE is being increasingly recommended as an indicator of recent spills because of its high mobility in the subsurface environment. Various states have established interim drinking water standards for MTBE ranging from 5 to 100 µg/L.[10]

4.4.2.3 Method Description and Selection

In order to determine whether a release from an underground tank has occurred, it is necessary to select analytical methods that will accurately identify, if not quantify, key components of the stored product. However, some controversy exists regarding which analytical methods are "best"; the conflict between analytical costs and data requirements emphasizes the need to develop an efficient and well-thought-out analytical protocol. This section describes the particular application of each method and provides some recommendations for assembling an informative analytical program.

For petroleum products, the choice of analytical method depends somewhat on the age and composition of the spill. Also important is that petroleum products are not always chemically distinct; the lighter components of fuel oil, for instance, consist of the same compounds as the heavier ends of gasoline, and usually contain small amounts of BTEX as well.[2,10] Since many hydrocarbons degrade in the environment to some extent, the composition of released material will change over time. These factors point to the need for more than one analysis to detect the presence of a petroleum release.

To ensure that the results are legally defensible and as consistent as possible, only EPA-or State-approved methods should be specified. Commonly used EPA methods are contained in the documents specified below.

- "Methods for Chemical Analysis of Water and Wastes" EPA Manual 600/4-79-020 (1983) (Methods E100 through E400 Series)
- "Methods for the Determination of Organic Compounds in Finished Drinking Water and Source Water" U.S. EPA, Environmental Monitoring and Support Laboratory (September, 1986) (E500 Series Methods)
- "Methods for Organic Chemical Analysis of Municipal and Industrial Wastewater" 40 CFR 136, Appendix A (1985) U.S. EPA (E600 Series Methods)
- "Test Methods for Evaluating Solid Waste" SW-846, 3rd Ed. (1986) U.S. EPA (SW Series Methods, for water and soils)

Of these, the "SW" (solid waste) series methods are normally specified for investigations conducted under the Resource Conservation and Recovery Act (RCRA) or the Comprehensive Environmental Response, Compensation, and Liability Act (CERCLA) regulations. Exception analyses are the analysis for Total Petroleum Hydrocarbons, E418.1, as well as for pH, total dissolved solids, common anions, and other standard groundwater evaluations.

The following procedures are normally used to analyze soil and water for petroleum products. Since most of these methods were not developed specifically for petroleum analysis, some limitations exist with some of them that need to be accounted for during subsequent data evaluation.

4.4.2.3.1 *Method Descriptions*

Method E418.1: Total Petroleum Hydrocarbons (TPH) — This method uses infrared absorbance to determine the quantity of total recoverable hydrocarbons (both aromatics and aliphatics) in water, wastewater, and soils. Petroleum is extracted from the sample using either a fluorocarbon (Freon 113) or, for soils, sonication (method SW 3550). This method does not distinguish between hydrocarbon types, but rather produces a single value for all petroleum hydrocarbons combined. For this reason it is considered a general indicator only; more detailed information about compound identity must be obtained for remediation.

Concerns with this method revolve around the loss of some petroleum fractions during the extraction procedure and the use of accurate standards. With a freon extraction, as much as 90% of the shorter hydrocarbons may evaporate, while heavier ends do not extract well.[10,12] Likewise, the agitation that results from sonication can cause the loss of a significant amount of volatile components.

In addition, acquiring laboratory standards (to calibrate the instrument) that closely resemble the sample is difficult because of the highly variable nature of fresh and weathered petroleum. As a result, precision and accuracy of the final analytical results tend to be lower for this method than for others, affecting somewhat the comparability between sample rounds.[10,13]

Nonetheless, E418.1 remains one of the most common methods for petroleum analysis currently in use, being a required procedure for UST site assessments in over 40 states.[14] Relatively inexpensive, this method does not require highly trained analysts and can be performed within a fairly short turnaround time by most laboratories.[10] E418.1 provides an indication of the general amount of hydrocarbons present at a site, and is a useful supplement to the more quantitative, but less comprehensive, gas chromatography methods described below.

Method 8010: Halogenated Volatile Organics (GC/PID) — This gas chromatography/photoionization detector (GC/PID) method determines low levels of 44 halogenated organic compounds in both soil and water.[11] Halogenated compounds are those containing chlorine, fluorine, and other "halogens," and include such common solvents as trichloroethane, trichloroethene, and carbon tetrachloride.

Method SW 8015: Nonhalogenated Volatile Organics (GC/PID) — This method uses a gas chromatograph and photoionization detector to determine low levels of six nonhalogenated volatile organics in both soil and water.[11]

These compounds include acrylamide, diethyl ether, ethanol, methyl ethyl ketone (MEK), methyl isobutyl ketone (MIBK), and paraldehyde.

Method SW 8020: Aromatic Volatile Organics (GC/PID) — This method uses a gas chromatograph and photoionization detector to determine low levels of aromatic compounds in both soil and water. These analytes include benzene, toluene, ethylbenzene, xylenes, chlorobenzene, 1,2-dichlorobenzene, 1,3-dichlorobenzene, and 1,4-dichlorobenzene. This method has a limited capacity to detect aliphatic compounds or MTBE.

Method SW 8240: Volatile Organic Compounds (GC/MS) — This broad-spectrum analytical method employs a gas chromatograph followed by a mass spectrometer (MS) to identify and quantify 45 halogenated and nonhalogenated organic compounds in water and soil. Although more time consuming and considerably more expensive than the GC/PID methods described above, this method allows precise identification of a wide range of common organics. Because of compound loss during the operation of the mass spectrometer, however, detection limits for this method are approximately 10 times higher than GC/PID methods for aqueous samples.[11]

Method SW8270: Semivolatile Organic Compounds(Base/Neutral and Acid Extractables)(GC/MS) — This GC/MS method identifies and quantifies semivolatile and nonvolatile organic compounds, including phenol, phthalates, and naphthalene. This method will detect some, but not all, of constituents typical of the middle distillate range. Identification of all target compounds in soil samples with this method is sometimes difficult, however, since different compounds will merge into a single peak or hump on the GC printout.[10] This difficulty is less problematic with water samples, since some group separation occurs when the product dissolves in water; the aqueous phase generally favors lower molecular weight hydrocarbons.[10]

4.4.2.3.2 *Method Selection*

The choice between using the methods SW 8010, 8015, and 8020, which utilize a selective detector such as a PID, and the more expensive SW 8240, which employs a MS for compound identification, depends on the use of the data and the detection limit required. Where the composition or likely contaminants of a sample are largely unknown, and the goal of the analysis is to identify constituents of concern for subsequent investigations or remediation, then SW 8240, the GC/MS method, is recommended.[15] The MS provides more specific identification of compounds and yields fewer false positives than selective detectors. However, as mentioned earlier, the mass spectrometer yields higher detection limits than GC/PID methods, with detection limits of 5 to 10 μg/L for aqueous samples, as opposed to 0.2 to 1 μg/L for GC/PID methods.[11]

The methods using selective detectors are most useful after the contaminants have already been identified, and when sampling is for confirmatory or monitoring purposes only.[15] These methods can achieve the very low detection limits necessary for the comparison of groundwater data to drinking water standards. Detection limits for aqueous samples from the three methods range from 0.03 to 1 µg/L. Detection limits for soil are comparable to method SW 8240, though the limit for SW 8015 is 1 mg/kg.[11]

Thus, to detect residuals from a tank whose contents, both current and historical, are known, and particularly if fumes are detectable either by monitoring equipment or olfactory senses, a GC/PID method (SW 8010, 8015, or 8020, depending on the tank contents) coupled with E418.1, Total Petroleum Hydrocarbons (TPH), would be a good choice. The presence of fumes indicates that significant amounts of volatiles are present in the soil, which can be detected by the GC/PID methods, and the TPH analysis would give a general assessment of the quantity of heavier petroleum components.

TPH data is particularly useful for highly contaminated samples, most of which can be analyzed directly by infrared absorption instruments. Because of the sensitivity of GC, however, highly contaminated samples destined for GC analysis must be diluted, which may result in detection limits of several hundred mg/kg and thus render "nondetect" values fairly useless. TPH and/or BTEX analysis are the analyses most frequently recommended by states for the analysis of soils potentially contaminated by petroleum.[1,16] In general, TPH is a helpful and relatively inexpensive analysis that is frequently recommended for contaminated fuel sites.[1-3,15]

MTBE, the highly mobile additive in gasoline, can be detected by method SW 8020, but must be specifically requested as a target compound. It is not a normal analyte of this method.

In cases where tank contents are currently or historically unknown, where a release may have occurred in the past, or where contamination from another source is suspected, the more precise identification abilities and broader range of a GC/MS method is warranted. The composition of weathered petroleum products, especially volatile ones like gasoline, can change considerably over time as the proportion of volatile components diminishes; analyzing for only the volatiles targeted in GC/PID methods may overlook significant quantities of heavier residuals. To identify both the remaining volatiles as well as the more persistent and less mobile heavy components, both SW 8240 (volatile organics) and SW 8270 (semivolatiles) are appropriate.

Basing management decisions only on the concentrations of relatively transient BTEX has distinct pitfalls, as illustrated by the following account of the remediation of a gasoline tank leak:[10]

In response to the leak, the tank was removed, free product recovered, and a "packed tower" air-stripping device installed to treat contaminated groundwater. Over several years of operation the influent of the "packed tower" was routinely monitored for "volatiles" using U.S. EPA methods 602 and 624 (*author note:*

these are equivalent to SW 8020 and 8240). When the concentration of benzene, toluene, and related compounds in the groundwater had fallen below detection limits of 2 μg/L it was concluded that the established objectives of the remediation program had been met. The water treatment equipment was dismantled and removed from the site.

Several years later, a sample from a monitoring well on the site confirmed that the concentration of the "volatile" hydrocarbons in the groundwater was close to the detection limit. However, this was not the case for the "extractable" hydrocarbons. The total concentration of these compounds was in the parts per million range. The initial response to these data was that a fuel oil leak was responsible. This was a reasonable conclusion based on the "volatiles"/"extractable" analytical scheme. However, it did not take into account the fact that many of the more water-soluble compounds found in middle distillate fuel oil are represented in the "heavier ends" of gasolines. Ultimately, it was determined that there had been no recent releases of fuel oils on the site. The compounds that had been detected in the groundwater were gasoline residues from the original spill.

Further remediation is under consideration and it is not surprising that there are several legal and financial complications.

Tetraethyl lead levels can be determined by methods SW 7421 (water) and SW 7420 (soil). As discussed earlier, these methods do not distinguish between naturally occurring inorganic lead and gasoline-derived organic lead, so a background sample may need to be collected.

Ethylene dibromide can be determined in water by method E504.

SW 8270 should be specified for suspected releases of middle distillates such as diesel fuel, since it targets some, but not all, of the compounds commonly found in these products. This may be supplemented by a product-specific GC/PID method or SW 8240, depending on whether the identity of the compound is known.

In reality, the choice of analytical protocols is often constrained by the size of the tank owner's bank account. However, while fewer analyses are, of course, less costly, they also entail a larger margin of error and a greater potential for missing contamination; these equally important realities need to be understood by both the tank owner and consultant.

If only a single analysis is to be performed under a petroleum tank, then TPH (E418.1) is a good choice. For gasoline tanks, supplementing this with a BTEX analysis is also advisable.

Full characterization of sites containing unknown products or chemically diverse products such as waste oil, require a broad range of analyses targeting the specific compounds of concern. A possible sequence for this case is as follows:

Soil	E418.1	Total Petroleum Hydrocarbons
	SW 8240	Volatile Organic Compounds
	SW 8270	Semivolatile Organic Compounds
	SW 6010	Metals Screen: RCRA Metals
	Mercury	Sw 7471

For water samples, the sequence might be the following:

Water	Field tests	Conductivity, pH, Temperature
	SW 8240	Volatile Organic Compounds
	SW 8270	Semivolatile Organic Compounds
	SW 6010	Metals Screen: RCRA Metals (except lead)
	SW 7060	Arsenic
	SW 7421	Lead (this method has a lower detection limit than SW 6010)
	SW 7470	Mercury
	SW 7740	Selenium

Metals analysis of water samples should include both the total and dissolved component to distinguish between metals dissolved in the water and those associated with suspended particles.

Metal analyses are important because heavy metals are frequently associated with engine and metal-related industries that have a high use of solvents and fuels. Waste oil or used solvents can contain lead or manganese from fuel residues; lead, cadmium, chromium, or zinc from painted or plated surfaces or stainless steel; or small quantities of the numerous other metals widely used throughout industry. In addition, waste oil tanks frequently become the repository for waste chemicals in general, and thus may contain chlorinated hydrocarbons from carburetor cleaner or other unknown chemicals.

Some states and federal agencies have their own recommended analytical strategies for petroleum tank sites. California guidance, for instance, recommends that gasoline tank excavations be sampled for both BTEX and TPH, with lead and EDB included if site specific conditions warrant.[3]

For nonpetroleum products, the analytical protocol is, in some ways, more straightforward. In each case, an analytical procedure is specified that targets, as nearly as possible, the contents stored in the tank. For acids and caustics, pH is the obvious choice, although pH can be altered by natural soil and groundwater buffers. Depending on the use of the tank, a metals screen may be useful to identify metals present in acid solutions.

4.5 FACTORS AFFECTING SAMPLING RESULTS

Data interpretation is arguably the part of site assessment that requires the most scientific expertise. Because of the complexity of subsurface conditions at many sites and the many components of petroleum and other products, accurately evaluating subsurface conditions may involve considerably more time and energy than simply noting whether reported values fall above or below regulatory limits.

Although a comprehensive treatment of data interpretation is beyond the scope of this book, the following points should be kept in mind when drawing conclusions about site conditions.

1. **Soil composition** — Because different soil types have different physical and chemical properties, the type of soil under a site can affect the location and distribution of petroleum and other compounds. The concentration of contaminants in the unsaturated zone above the water table depends on the soil sorptive capacity and soil capillarity, which is in turn related to grain size; these characteristics influence the extent to which contaminants can be held on and between grains. In general, clean, coarse sand, which has a low sorptive capacity and capillarity, will hold relatively little contamination between grains. Tighter soil with a large clay component will have higher capillary action and relatively high retention of contaminants.[17] Sandy soil has been observed to leach more BTEX from petroleum products than clay soils.[18]

2. **Composition of petroleum products** — As gasoline and other petroleum products age in the soil environment, their compositions change due to the volatilization, biodegradation, and solution of the BTEX and other light components. Biodegradation rates are enhanced in sandy soil or aquifers where the amount of dissolved oxygen is high; in one groundwater study, natural aerobic biodegradation was the major mechanism responsible for the soluble benzene reduction, with biodegradation rates approaching 1% per day.[19,20] In fresh gasoline spills, the volatile component may be present in water in a concentration one order of magnitude higher than semivolatile components, but this relationship changes greatly with weathering.[10]

   Also, as noted earlier, light and middle distillate products are not chemically separate entities, but have many of the same compounds in different quantities. The heavier ends of gasoline consist of the same compounds as the lighter ends of middle distillates, while significant amounts of BTEX can leach from middle distillate products such as diesel fuel.[2,10,21] In addition, BTEX is present in many products, not only petroleum hydrocarbons; thus the simple presence of BTEX alone, in the absence of other evidence, is not necessarily an indicator of petroleum contamination.[2]

   Thus, low BTEX values coupled with, for instance, high TPH values, may represent either an old gasoline spill, from which the volatile fraction has disappeared, or a heavier product, such as diesel or kerosene. Unless gasoline is known to be the only possible source, further analyses will be required to identify the residual. This is particularly true on sites where several sources may exist in close proximity, such as a service station with separate USTs for leaded and unleaded gasoline, diesel fuel, kerosene, waste oil, and heating oil.

3. **Other Sources** — It is important to keep in mind that data gaps are inherent in working with old tanks. High BTEX values from the site of a tank abandoned years ago, for instance, may actually represent the migrating front of a relatively fresh fuel spill at an adjoining site, where an existing tank is leaking or is routinely overfilled. Similarly, high values of semivolatile components around a gasoline tank may reflect residuals from an old spill or from a nearby heating oil or diesel tank. Because data from old sites containing abandoned tanks are especially liable to be absent or incorrect, in these cases it is particularly important to avoid the investigative trap of searching only for what one expects to find.

4. **Analytical Limitations** — As discussed under the description of analytical methods provided earlier, many methods have limitations that affect the strength of the conclusions based on them. Since extractions are often inefficient, especially when the soil contains high quantities of organic matter or clay, it is best to be conservative when defining the extent of contamination.[12]

5. **Location of Samples** — Surface or subsurface soil sampling may easily miss contamination occurring at greater depths. In addition, some chemicals cause a significant increase in soil conductivity, and thus go to great depths without leaving significant amounts of contamination in the unsaturated zone.[22] Low data values should be examined in light of factors such as site geology and tank history, and other factors before being accepted as an accurate reflection of site conditions.

6. **Background Soil Concentrations** — Taking a sample of "clean" soil from the site to establish background conditions is often helpful. Benzene, for instance, may occur naturally in some soils at concentrations of 1 to 5 mg/kg,[22] and some California soils are naturally high in lead. Identifying these phenomena can help isolate the effects of tank releases.

4.6 HOW CLEAN IS CLEAN?

After analytical results have been received from the laboratory, the consultant (or whoever else) conducting the site assessment must determine whether the tank hole is "clean" or otherwise free, or free enough, of contamination. Unfortunately, no federal standards exist for concentrations of petroleum products or most other constituents in soil, and only two states — Alabama and Montana — had formal soil cleanup standards for petroleum at the time of this writing.[14]

However, many states, especially those states that require soil sampling, do have informal standards. For TPH, 100 mg/kg (ppm) is commonly used, but in many locations the limit is set on a case-by-case basis, so may be altered upward in urban areas with no groundwater resources, or downward if the site is in an environmentally sensitive area; however, the determination is up to the regulators. The value of the informal standard should be identified before the site assessment is completed.

For inorganic constituents, the consultant or state regulator may use other criteria to determine when the site is "clean". These include (1) concentrations in a "background" soil sample collected from a "clean" part of the site; (2) concentrations "naturally" present in soil, as determined by a review of the scientific literature; (3) the detection level of the analytical technique; (4) a site-specific value developed from a risk assessment of the constituent, or (5) proposed EPA RCRA action levels for corrective actions at Solid Waste Management Units (551 FR 30798). The first two options are useful for naturally occurring constituents, such as lead or chromium. The last two are often used for toxic compounds that have no natural source, and which may present a significant public health hazard.

Groundwater samples are usually compared to the Maximum Contaminant Levels allowed for drinking water, as specified in 40 CFR 141.11 and 141.61. In some so-called "nonattainment" areas, higher values may be allowed, but these values are normally negotiated with the state regulators.

As noted earlier, some states do not require sampling at all: a tank hole or excavated soil that is free of discernable odors or visual contamination is considered "clean" for state purposes. While this method is undoubtedly useful and even accurate in many cases, it is often the case that soil will have significant contamination (e.g., hundreds of parts per million) and yet show no obvious sign. This fact can put the consultant conducting the site assessment in an awkward position: he or she may feel professionally obliged to (1) recommend sampling of the excavation; and (2) recommend remediation of a site that the state has ruled "clean". Understandably, tank owners are sometimes suspicious of these recommendations, especially when the consultant would be performing the work for an additional fee.

However, additional or confirmatory sampling and, if necessary, remediation, are often in the best interests of the tank owner in order to certify the condition of the site at the time of sale. An increasing number of banks and other financial institutions are requiring environmental assessments of commercial and industrial properties before advancing money for renovations or purchases. These assessments naturally zero in on USTs as excellent candidates for contamination: if there has been a removal, especially of old tanks, proof may be required to show that contaminated soil does not remain on site. Few consultants or experienced loan officers will consider anything but actual soil samples as proof, and in consequence may recommend or require test pits or soil borings to verify subsurface conditions, even though the state approved the site as clean. If contamination is discovered, chances are that the loan will be refused unless the site is remediated, which can be financially disastrous if new tanks or other structures have been installed or erected on the site. For this reason, the most prudent course of action is to conduct a thorough site investigation — which includes soil samples — at the time of removal, and to remediate as necessary when the holes are open. Remediation at this point may only involve another 20 minutes of excavation; remediation later may mean exhuming tens of thousands of dollars worth of equipment or spending equal amounts for an *in situ* remediation system.

4.7 EQUIPMENT DECONTAMINATION

As a standard component of the field program, all sampling equipment must be decontaminated to ensure that residual components from equipment handling or previous sampling do not distort the results of subsequent samples. Contamination by sampling devices and material can contribute relatively large errors in comparison to analytical procedures, especially when analytes of interest are at low concentrations.[5] Recommended procedures for decontamination vary somewhat according to the nature of the contaminant; a comprehensive decontamination scheme is illustrated below.[4,23]

1. Soap and tapwater wash
2. Tapwater rinse
3. Rinse with distilled water (ASTM Type I or II)
4. Rinse with spectrographic grade acetone or methanol (for petroleum residuals)
5. Rinse with spectrographic grade methylene chloride or hexane (for PCBs)
6. Air or flame dry

Sampling equipment should be of stainless steel, and should be decontaminated (1) before collection of the first sample, and (2) between samples that are collected for individual analysis, and (3) after sampling is complete. Decontaminating equipment between sampling points is not necessary if the collected soil will be combined into a composite sample.

Likewise, sampling containers must also be cleaned according to a similar procedure. Container decontamination is normally carried out by the container supplier, which is often the analytical laboratory.

4.8 FIELD SCREENING METHODS

4.8.1 Uses and Limitations of Field Screening Techniques

Field screening methods include the use of portable, on-site field monitoring instruments to obtain a relatively rapid determination of the presence or relative levels of site contamination without direct soil or water sampling and laboratory analysis. Field screening is particularly attractive during tank removals, since potential petroleum contamination can be assessed while the heavy equipment is still on site to remove contaminated soil; similarly, the rapid return of site information can be used to direct and refocus a soil boring program while the fieldwork is still in progress. In contrast, laboratory analyses, while generally providing high quality data, have the dual disadvantages of high cost and relatively long turnaround times, both of which can limit the progress and scope of site investigations. For these reasons, relatively fast and low-cost field screening techniques have become increasingly popular as initial site investigation activities.

Field screening methods are particularly useful for the following:

1. To detect quickly the presence and relative levels of contamination at a site
2. To select samples for laboratory analysis from among many soil samples collected at a site
3. To identify appropriate sites for soil borings and monitoring wells

Field screening can provide an initial assessment of certain kinds of contamination (e.g., petroleum hydrocarbons) so that subsequent sampling can focus on known areas of concern, thus minimizing the overall cost of the investigation. While exact contaminant concentrations can rarely be obtained in the field, relative levels can be determined that nonetheless reveal "hotspots", plume boundaries, and other important contamination characteristics.

Field screening methods do, however, have limitations. Some constituents, such as metals, are not detectable by vapor-sensing techniques. Many techniques rely on indirect measures of contamination (e.g., vapor production or changes in physical properties of soils), and so are less accurate than methods which detect and measure the constituent of concern directly. These methods provide information on relative levels of contamination, but the actual numbers obtained are rarely comparable to laboratory results. In addition, not all methods distinguish between constituents, but rather provide only a measure of the total effect of all constituents combined. And finally, because contamination is measured in place, subsurface conditions may interfere with the ability of remote instruments to detect contaminant presence. In some cases, the use of more than one screening technology may be necessary to obtain an accurate picture of contaminant distribution.

Field screening methods are normally used in advance of, or in conjunction with, soil sampling efforts. When contamination has been confirmed, field screening methods may find further use in identifying the most appropriate location for monitoring wells or soil borings, the conventional but relatively expensive backbone of most remedial investigations.

The sections below describe key aspects of two of the more common field screening techniques — vapor analyses and geophysical survey. Because other methods exist or are under development, however, this treatment should not be considered exhaustive.

4.8.2 Vapor Analyses

Vapor analyses broadly include those techniques which collect and analyze vapors emitted from soil or groundwater. By necessity, the use of this technology is limited to those constituents which are volatile to some extent at normal atmospheric temperatures; these include many organic compounds but exclude metals. Thus, vapor analysis has found wide application in the investigation of petroleum releases, particularly those involving gasoline, diesel, chlorinated solvents, or other relatively volatile compounds. Heavier grade petroleum products, such as No. 6 fuel oil, are less detectable by this method.

Screening based on vapor detection can be divided into two general types: headspace analysis of individual soil samples, and evaluation of on-site soil conditions by soil gas surveys.

4.8.2.1 Vapor Detection Instrumentation

Field vapor detection instruments are normally compact, relatively rugged devices that either draw air into the internal chamber of the instrument or require that an air sample be injected into a special port. While there are several manufacturers of these instruments, the instruments themselves are of four different types: photoionization detector (PID), flame ionization detector (FID), portable gas chromatographs (GC), and colorimetric tubes.

A PID measures the number of molecules that lose electrons, or ionize, when drawn through the beam of ultraviolet (UV) light. The freed ions produce a current proportional to the number of ions present, and the current flow is quantified by the deflection of a dial needle on the front of the instrument.

The ability of the PID to detect a given compound depends on the compound's "ionization potential" or the amount of energy, expressed in electron-volts (eV), required to break the electron free. Thus, a PID equipped with a 10.2-eV lamp will ionize all compounds with ionization potentials of 10.2 eV or less, without distinguishing between them. Likewise, an 11.7-eV lamp will detect all compounds with an ionization potential of 11.7 eV or less, including those ionized by the 10.2-eV lamp. Lamps are interchangeable, so it is important to ensure that the target compound, for instance benzene or tetrachloroethylene, has an ionization potential less than the lamp intended to detect it.

PIDs are best suited to the detection of ring compounds, but cannot distinguish one aromatic from another. They do not detect methane, but some models will detect water, which limits their usefulness on humid or rainy days.

PIDs are probably the most widely used vapor monitoring instruments because of their small size, light weight, and ease of calibration and use. Common PID models include the HNu (Models P-101, HW-101, and IS-101), the Photovac MicroTip, and the Thermo Environmental Instruments Organic Vapor Meter (OVM).

Flame ionization detectors (FIDs) are similar to PIDs except that they use a hydrogen flame, rather than UV light, to ionize compounds. Sensitivities are similar for aromatic and aliphatic compounds, and at least one model can be operated either to detect or to not detect methane. They are generally less sensitive to high humidity than PIDs, but, like PIDs, measure total organic vapors, not specific constituents.[24]

One commonly used field FID is the Foxboro Century Organic Vapor Analyzer (OVA) Model 128.

Gas chromatographs (GC) are the primary method used in laboratories for the identification of organic compounds, and portable GCs bring some of the advantages of the GC into the field. GCs have the ability to identify compounds (with certain limitations) as well as quantify them, but require skilled personnel to calibrate and operate the instruments.

The operating principles of the GC are as follows. Sample gas, or gas produced by the vaporization of a liquid sample, is passed through an absorbent column, where the different components of the gas are absorbed and desorbed according to the unique chemical, physical, and structural properties of each component. As a result, each component elutes from the column at a different and compound-specific period of time, which is referred to as the retention time for that compound. The GC is in turn coupled with a PID, FID, or MS which registers and quantifies each compound passing through and produces the information on a linear chart of successive peaks, where each peak represents the presence of a compound. Compounds are identified by their retention time and quantified by integrating the area under the peak. Identification is further aided by the use of reference charts for known compounds.

GCs are relatively sensitive instruments that are affected by changes in temperature and other environmental factors. Ideal conditions for using GC are dry weather with small temperature changes, but in all cases the instrument should be calibrated in the conditions under which it will be used. Due to the relative complexity of calibration and operation of the instrument, as well as the need for informed interpretation of results, GCs require the use of trained personnel.[25]

Under ideal conditions, field GCs can produce analytical results generally comparable to those obtained in the laboratory. For a number of reasons, however, field GCs have been found to occasionally produce false positives when compared to the GC/MS procedure of EPA method SW 8240.[26]

GCs are produced by a number of manufacturers. The Foxboro OVA Model 128, described above, has a GC built into the unit.

Colorimetric tubes contain a compound-specific (or compound-type specific) absorbent which changes color when exposed to the target compound. A measured amount of the gas or atmosphere is drawn through the tube containing the absorbent, and the resulting change in color is compared to a standard color chart. The length of absorbent colored by the sample is proportional to the concentration of the target compound.

Colorimetric tubes are inexpensive, simple, and quick to use, but require knowledge of the target compound ahead of time so as to select the appropriate tube. In addition, the tubes provide data representative of one place and time only; they do not have the scanning capabilities of continuous monitoring devices like the PID and FID. Nonetheless, they provide an inexpensive and useful means of generally identifying (through tube selection) and quantifying specific compounds or groups of compounds, and can be a useful supplement to the total organic concentration information provided by the PID and FID.

Cross-sensitivity (i.e., coloration of the tube by something other than the target compound) can interfere with accurate readings of the tube. One tube model designed for toluene, for instance, will read anomalously high if benzene is also present. In addition, the tubes can become saturated during periods of high humidity, yielding low results of the target compounds. Finally, imprecise boundaries between colored and noncolored portions of the tube can sometimes make precise interpretation difficult.[25]

A common brand of colorimetric tube widely used in remedial investigations is the Drager Tube. The tube is fitted to a hand-held pump, which is squeezed a specified number of times to draw the appropriate amount of sample air through the tube.

4.8.2.2 Headspace Analyses

Headspace analysis is the evaluation of the vapors emitted by a container-ized soil sample, or in its simplest form, of the vapors released by any discrete sample of soil. The usual goal of headspace screening is to select samples for laboratory analysis when a large number of samples are or could be collected, for instance during continuous sampling in soil borings; however, other field uses include evaluating soil for disposal, locating "hot spots" in exposed soil, and selecting locations for further study. Samples can be analyzed by either a PID, FID, or field GC; field GCs take somewhat longer to operate, but all will provide results in less than 15 minutes. A range of practices for the assessment of headspace or surface vapors exists.

4.8.2.2.1 *Surface Vapor Screening*

In its simplest form, surface vapor screening of soil consists of holding the probe of a PID or FID over an exposed soil sample or soil location. Commonly used to evaluate the soil contents of split spoon samplers or to assess excavated soil quickly, this method is fast and most effective for detecting high concen-trations of volatiles released immediately from the soil matrix. Relatively low concentrations or quantities of compounds with relatively low volatility may not be detected by this method, nor may volatiles present around quantities of clay or other fine-grained soil.[27] In the field, cross-breezes or other air move-ment around the soil surface may dilute the air collected by the field instru-ment, resulting in an anomalously low reading. Nonetheless, this method is quick to perform and provides immediate results, important factors when excavation is underway. Surface vapor screening is widely used as an initial screening method during site assessments and tank removals.

4.8.2.2.2 *Static Headspace Analysis*

True headspace analysis involves containerizing the soil sample and mea-suring the headspace over the sample after a predetermined amount of time. One common method involves sealing the top of the glass sample collection jar with aluminum foil, waiting a specified period of time, and then poking the probe of the field instrument through the aluminum foil and noting the highest reading detected.[25,28] The effectiveness of this technique depends upon the consistency of the procedure, which should be in accord with the following steps.[25,28]

1. Soil should be added so that approximately equal volumes of vacant headspace are present in each jar, which are then sealed immediately.
2. Sample jars should be stored out of the sun, at a constant ambient or room temperature for a consistent amount of time, which may range from 5 to 10 minutes to 1 to 2 hours (it is critical that the temperature of each sample from the same site is the same prior to analysis).

3. The aluminum foil is then pierced with the probe of a PID or other instrument, and a reading taken as soon as the instrument needle has stabilized or whenever the highest reading is obtained. Drawing air out for prolonged periods may affect the vapor equilibrium in the sample container.

Preliminary experiments indicate that this method is capable of detecting less than 3 mg/kg of fresh gasoline in the test soil, and even lower concentrations if the sample is agitated to break up the soil and aid volatilization.[28] High concentrations of gasoline vapor tend to cause "quenching" of a PID, which occurs when hydrocarbon vapors are so dense that they act as an insulator in the PID's ionization chamber, reducing or eliminating the unit's response.[28] While preliminary comparisons of this method to the results of laboratory analysis did not always show good correlation, the method has been successfully used to detect petroleum releases and map groundwater plumes at contaminated UST sites.[28]

Concerns with this and similar methods focus on a number of issues.[7] Changes in flow rates resulting from pumping a sealed area may affect instrument readings, while air leaks around the aluminum foil seal may dilute the vapor sample. Changes in temperature and barometric pressure will affect volatilization rates, making comparisons between samples collected on different days problematic. Likewise, differences in soil lithology and moisture content between soil samples will affect volatilization rates; fine-grained, moist soils are likely to have a lower volatilization rate than coarse, dry ones. Also, soil may have entrapped pockets of gas which, when exposed and detected by the PID or FID, result in a false positive reading relative to measurement of the soil itself.[27]

4.8.2.2.3 Dynamic Headspace Analysis

Dynamic headspace analyses differ from static headspace analyses in that the sample container is agitated after the soil or water is added, rather than letting the sample sit and reach ambient temperature.[25] The agitation provides a large surface area for volatilization and helps breaks up clods of sample soil. Samples are typically agitated anywhere from 15 seconds to 2 minutes; the amount of time is considered less important than the necessity for the same time to be used for all samples at a site. If the ambient temperature is below freezing, agitation and analysis are performed in a heated area.[25]

Samples are analyzed either immediately following agitation or after a specific and consistent amount of time has elapsed. Vapor is sampled by pushing a probe through the tinfoil cover and withdrawing vapor. For GC analysis, vapor is withdrawn through a syringe inserted through the tinfoil.[25]

Although agitation helps release volatile components, this method is still subject to the same sources of error as the static headspace method described above. A more refined version of the dynamic headspace method has been developed that uses a polyethylene bag as the sample container. The general procedure used for this method is as follows.[25,29,30]

1. A 1-qt resealable (Ziploc®) bag is fitted with a three-way valve, one opening of which protrudes into the bag through a hole made approximately 2 inches down from the top. The bag is sealed to the fitting with gaskets.
2. Using the valve, the bag is then inflated until just taut by means of a handpump. The inflated bag is inspected for leaks and then deflated.
3. For soil analyses, 100 ml of distilled water is added to the bag, to which is then added 25 g (or some other consistent amount) of sample soil. For water analyses, 100 to 300 ml of sample water is added to the sample bag. After the soil or water sample is in place, the bag is then sealed and inflated until just taut and the valve to the bag is closed.
4. The sample is then agitated for 4 minutes by either a gentle rocking motion or the use of a magnetic stirrer.
5. Following agitation, the valve is opened and a sample of the headspace in the bag routed to a field instrument through a length of Tygon™ tubing. Securing the bag and valve on a ring stand during this step helps to stabilize the bag and prevent the accidental introduction of water into the screening instrument.

In order to evaluate the actual level of contaminants in the soil based on vapor concentrations, the procedure must be calibrated to the target compound or compound mixture. For a single constituent, dilution of an aqueous standard may be performed to calibrate headspace concentration with water concentration. Performing this procedure also permits testing field performance of personnel and equipment.[25,30]

For multiple constituents, four calibration methods have been suggested, as described below:[30]

1. A single component calibration may be performed and multi-component results reported as concentration equivalents.
2. A multicomponent standard having the same constituents in a similar proportion as that of the ground water may be used to generate a calibration curve.
3. A sample of the groundwater may be serially diluted to develop a relative concentration calibration curve, which can be later semiquantified by a laboratory analysis of the sample.
4. Headspace analyses of groundwater samples from wells where the concentration of constituents has already been determined by laboratory analyses may be used to generate a calibration curve.

Like other methods, headspace analysis is susceptible to various influences. Salinity and dissolved organic content of water samples can affect field standards, resulting in the generation of misleading calibration curves. Nonlinear instrument responses may be produced in a multicomponent sample, due to variations in individual concentrations in the mixture. Soil sample results may also be affected by sorption-dissolution effects at the soil-water interface, resulting in a lower total reading.[25]

With this and other methods using field instruments, anomalously high readings may be obtained if the field instrument detects compounds that are not targeted by the laboratory method used to confirm the field results.[27]

4.8.2.3 Soil Gas Surveys

The sampling and analysis of soil gas *in situ* (here used meaning "in place") is emerging as a popular technique for locating near-surface sources of petroleum contamination, and for obtaining additional information about appropriate locations of monitoring wells and soil borings. Because the techniques involved require little drilling or construction, soil gas surveys can provide a relatively large amount of information about a site at a fraction of the cost and time of conventional groundwater investigations, and thus constitute an important screening tool for initial assessments of tank sites.

Soil gas is the air that exists in the pore spaces of the soil above the groundwater, an area referred to as the vadose zone. Soil gas around soils or groundwater contaminated with volatile organic compounds, such as many components of gasoline, may contain "fumes" or detectable quantities of those volatile compounds, which are carried with the soil gas as it diffuses through the soil in response to soil conditions and chemical/physical gradients. Generally speaking, the more volatile a compound, or the more readily it converts to the gaseous state at a given temperature and pressure, the more amenable it is to detection in soil gas surveys.[31]

Soil gas samples can be collected in two ways, referred to as grab (or active) and passive sampling. Grab sampling involves the relatively rapid collection of soil vapors; in dynamic grab sampling, soil gas is pumped out through a metal tube inserted in the ground, and the sample is collected from the moving stream of air. Static grab samples are collected from a quiescent soil gas sample. Grab samples are typically analyzed immediately, usually by field instrumentation or at a mobile or on-site laboratory containing a GC and other detection instruments. Several locations may be analyzed per day with this method.[32,33] If particularly high quality data are required, for instance, on a site containing multiple plumes, laboratory analysis of samples may be used instead of field analysis, although the use of laboratory analysis will delay the receipt of sampling results.

Passive sampling, on the other hand, is performed by inserting a tube containing a charcoal sorbant into the soil and leaving it in place for days or weeks at a time, allowing soil gas to diffuse naturally into the charcoal. After collection, the sample is analyzed in the same manner as a grab sample.

Both methods have their relative advantages and disadvantages, although dynamic grab sampling is more widely used. Grab sampling with on-site analysis is useful because data on subsurface conditions can be obtained within hours, allowing subsequent sampling to be tailored to the results.[34,35] When a PID or other instrument is used, results can be obtained at a relatively low cost. However, analysis of soil gas with a field GC requires trained personnel and an on-site laboratory or other suitably controlled environment where the GC can be operated; these requirements tend to increase start-up costs. In addition, soil perturbations from pumping tend to make results hard to reproduce. In general, grab sampling and on-site analysis is most useful for projects where the site is largely unknown and the soil gas survey is exploratory in nature and able to be altered in response to sampling results.[32,33]

Grab samples may also be analyzed at a commercial laboratory rather than at an on-site facility. The enhanced precision and accuracy usually obtainable with laboratory-grade instruments are particularly useful for identifying a wide range of soil gas components, which may be necessary to differentiate between plumes of different materials or ages.[36,37] However, the normally obtainable laboratory turnaround time, 1 to 4 weeks, means that the advantages of real-time analysis are lost; thus the soil gas survey must be completed in advance of subsequent site activities. Some companies use laboratory-grade instruments in their mobile laboratories, a practice which combines the advantages of field responsiveness and laboratory standards.

Passive sampling is best suited to relatively small projects where the results are not needed for several weeks. Some knowledge of the site is necessary to select appropriate sampling locations, but the sampling itself requires less highly trained personnel than does grab sampling. For these reasons, passive sampling tends to be somewhat cheaper than grab sampling.[32,33]

Because soil gas samples are relatively inexpensive when compared with conventional soil borings and monitoring wells, a typical soil gas survey may involve the collection of a large number of samples, often on a single day. This is particularly true if the survey is intended to chart groundwater contamination rather than to identify surface spill locations.

Because soil gas surveys depend on the diffusion of gaseous constituents, their use is limited to sites where other factors or conditions do not interfere with subsurface gas transport. Undisturbed, dry, coarse soils with a minimum of organic matter represent ideal conditions.[32,38] Underground features that might block or distort subsurface transport limit the technology's effectiveness. These include clay layers; perched groundwater; buried foundations; irrigated soils above the contaminant source that could block upward gas migration; or disturbed or backfilled soils, whose higher porosity can act as a conduit for gases.[22,32,36,39] Soil gas sampling in tight soils (silt-clay mixtures) is susceptible to sampling errors which underestimate the total amount of petroleum detected.[40]

Soil gas surveys are also best suited to detecting products that do not break down or biooxidize to low volatility substances in the soil. Some compounds that do biooxidize in this manner, (e.g., Jet A fuel, diesel fuel, and heating oil), may be evaluated by other means.[38] (see Section 4.4.2).

Interpreting soil gas survey results requires a good knowledge of subsurface fate and transport phenomena as well as site- and chemical-specific information. Consultants involved with soil vapor data interpretation will need to take into account the following factors, described generally below:

- Sampling depth — Because volatile organic compound (VOC) concentrations are highest near the source of contamination, sampling too near the surface may miss VOCs arising from contaminated groundwater or other sources several feet down. The most effective sampling depth may need to be determined at or near the beginning of the survey, especially if contaminated groundwater is a potential source.[31,33]

- Depth to groundwater — Shallow groundwater may result in steep VOC concentration gradients between the water table and the land surface, so that relatively large changes in VOC concentrations will be detected as the distance above the water table of the sampling point varies. Atmospheric effects, such as air temperature and wind velocity, have a relatively greater effect on near-surface soil samples.[38,41]
- Biooxidation — Many petroleum hydrocarbons biodegrade in the subsurface to other compounds.[31-33,42] Benzene, for instance, will biodegrade to certain straight chain hydrocarbons, so that the soil gas survey must focus on those secondary products rather than on an indicator like benzene alone.[39,42] The presence of benzene, on the other hand, is indicative of a recent release that has not yet completely biooxidized.[32] Aerobic biooxidation is accompanied by the production of carbon dioxide, and evaluating carbon dioxide levels may be a useful interpretive tool when the products of biooxidation are not themselves volatile.[31,42]
- Site stratigraphy — As mentioned earlier, low VOC concentrations in soil gas can be caused by perched water tables, clay layers, or any other factors that interfere with the diffusion of subsurface gas.[22,32,38,39]
- Alternative VOC sources — Unusually high VOC concentrations can be caused by leaking sewer or natural gas pipelines.[38] Analyzing for benzene or some other product-specific compound can be used to determine the source of VOCs.[38]

In general, soil gas surveying is a relatively fast and inexpensive means of obtaining preliminary information about a site, and may save thousands of dollars in drilling and sampling costs. Unlike geophysical methods, discussed next, soil gas surveying has the advantage of measuring directly a phase of the compounds of interest, rather than assessing secondary effects on soil properties. Soil gas surveying remains, however, a screening method only, and must be followed by groundwater and/or soil samples to confirm and define the limits of contamination.

4.8.3 Surface Geophysical Survey Techniques

Surface geophysical techniques are investigative procedures that are implemented at the ground surface (rather than from aircraft or within boreholes) to detect and quantify differences in the physical/chemical characteristics of the subsurface. These differences may arise from a variety of sources, ranging from natural stratigraphy to the presence of buried drums or contaminant plumes; thus these procedures are finding increasing application as a screening method for hazardous waste sites. Some of the uses include the following:[43]

- Mapping of natural hydrogeologic features, such as bedrock surfaces
- Mapping of conductive leachates and contaminant plumes

- Mapping of nonconductive plumes of organic contaminants
- Location and boundary definition of buried trenches, or other subsurface discontinuities
- Location and definition of buried metallic objects such as abandoned tanks, pipes, and utilities

On a small scale, geophysical techniques may be used to ensure that proposed drilling sites are not underlain by boulders, concrete pilings, or other structures that may damage the drill bits.

For the tank owner, the most likely use of geophysics may be to locate abandoned tanks and pipelines. However, in the event of contamination, additional methods may be recommended to help substantiate the findings of soil vapor studies of organic contaminants or to trace the boundaries of inorganic chemical plumes. When the released compound is heavier than water (e.g., chlorinated solvents), geophysics may be used to locate the lower confining layer over which the contaminant plume may be located.[43]

In general, geophysical techniques are most useful in homogenous soils or where stratigraphic boundaries are distinct. They are relatively fast to complete and provide immediate information about subsurface conditions, although subsequent data interpretation may be necessary to obtain a more accurate assessment. Some methods can be performed without disturbing asphalt or concrete surface paving.[44] Most techniques employ relatively sophisticated equipment and thus require trained and experienced personnel; in addition, because of the fairly narrow capabilities of each method, more than one technique may be necessary to characterize fully subsurface anomalies.[45,46]

4.8.3.1 Method Descriptions

Although a variety of geophysical techniques have been used at hazardous waste sites, only the more common are briefly discussed here. These include metal detection, magnetometry, ground penetrating radar (GPR), electromagnetics (EM), and resistivity.

4.8.3.1.1 *Metal Detection*

This method uses common commercial metal detectors to locate buried metallic items, such as drums or pipelines, or to trace out the location of utility trenches, which provide a route for contaminant migration. These detectors can locate any kind of conducting metal, including steel, aluminum, copper, and brass, and thus are widely used by treasure hunters and utility companies. Metal detectors are relatively inexpensive and easy to use, and information is obtained quickly and continuously.[43,47,48]

The effective penetrating depth of metal detectors increases with the size of the object detected. Buried drums may be detected at depths of 3 to 10 ft, while a large deposit of several drums may be detected at depths of 10 to 20 ft.[43]

4.8.3.1.2 *Magnetometry*

Magnetic surveys use a magnetometer to detect and measure local distur-
bances in the earth's magnetic field that exist as the result of buried ferrous
metal (iron and steel) objects. The magnetometer is particularly useful in
locating buried steel tanks and iron pipelines, particularly in areas where other
dissimilar metal types are present.[46,47] Because the magnitude of the detected
disturbance depends on the target object's size, shape, depth, orientation, and
susceptibility,[43] the observed response of similar objects may vary.

Depending on the terrain, up to 5 to 10 acres can usually be surveyed in a
single day[49] and single drums up to 20 ft deep can be detected. Quantities of
drums can be detected at depths of 65 ft or more.[43] Readings can either be
obtained continuously or at discrete intervals, depending on the type of magnetometer used.[43]

Interferences to magnetometer readings can be caused by iron oxides,
heterogenous soils, or cultural features such as power lines, substations, and
above- or below-ground structures. In addition, daily fluctuations in the earth's
magnetic field will distort background readings unless the survey is referenced
to an established base station every few hours.[47,49]

4.8.3.1.3 *Ground-Penetrating Radar (GPR)*

GPR uses high frequency radio waves to locate subsurface objects or
surfaces of sufficiently different electrical properties from the surrounding soil.
Energy emitted from an antenna at the surface is reflected back from subsurface objects and picked up by a receiving antenna, which detects variations in
the return signal. Because of the high frequency of energy used, the GPR
provides one of the highest levels of detail of all geophysical methods.[48]

GPR works best in highly resistive soils, since areas of high conductivity
(e.g., clay soils or conductive groundwater) tend to disperse the energy pulse.[48-
50] The best results are achieved in dry, sandy soil or rocky areas; however, good
penetration can be obtained in saturated soil if the conductivity of the pore fluid
is low.[43] Effective detection depth is thus highly site specific, but detection to
30 ft or more is common, and penetration to over 100 ft has been reported.[43,49]

GPR is useful for locating metallic and nonmetallic drums, for approxi-
mating drum density, for identifying the depth and location of buried trenches,
and for determining plume boundaries.[48] In addition, GPR can also be used
to map stratigraphic characteristics such as soil layers, depth to bedrock,
buried stream channels, rock fractures, and subsurface cavities. GPR has the
advantage of allowing continuous surveying and is unaffected by unrelated
sources of electromagnetism.[43,49] High resolution, ranging from inches to
several feet, can be achieved, depending on the frequency of energy used.[43]
However, the equipment is cumbersome, and must be driven or dragged over
the site surface.

4.8.3.1.4 *Electromagnetic Conductivity (EM)*

EM, also known as electromagnetics or terrain conductivity surveying, uses a low frequency electromagnetic field produced by an electric current to measure the electrical conductivity of subsurface soil, rock, groundwater, and target objects. This method is useful for mapping hydrogeology, conductive leachate plumes (primarily those containing ionic and inorganic constituents), and the boundaries of buried drum sites, trenches, and metallic utility lines.[48,49,51,52] Plumes of organic compounds can be mapped when they are of a size or density sufficient to reduce conductivity.[43] However, because this method does not differentiate between sources of conductivities, distinguishing between buried drums and a contaminated plume may in some cases be difficult.[48]

EM instrumentation is hand carried and is not affected by the ruggedness of the terrain.[49] It can be economically employed for shallow profiling of depths of 15 to 50 ft, although station-by-station profiling of 25 to 200 ft depths is possible under good conditions.[43,48] Stray sources of electromagnetism — for instance, steel utilities, pipes, cables, or fences — impair the effectiveness of the EM technique,[48,53] as do heterogenous soils with natural variations in conductivity.[54]

4.8.3.1.5 *Resistivity*

In a manner similar to conductivity surveys, resistivity techniques differentiate between subsurface strata or objects, based on their resistance to an applied electric current. A pair of electrodes is driven into the ground, an electrical current is applied, and the resulting potential field (voltage) is measured between a second pair of electrodes. Subsurface resistivity is then calculated from the geometry and separation distance between electrodes, the applied current, and the measured voltage. Since most soil and rock minerals are highly resistive, the measured resistivity applies primarily to pore fluids.[43]

Resistivity surveys may be used for either profiling (defining lateral changes in resistivity) or sounding (determining vertical changes).[43,51] Thus, these surveys are useful for determining natural components of the hydrogeological environment, for defining contaminant plumes, and for locating buried wastes.[43,47] Using computer methods for sounding, the depth and thicknesses of up to four subsurface layers may be identified.[43] Because of the need for electrodes, resistivity surveys are slower and more costly than EM methods, but provide more detailed data.[48]

Resistivity methods are useful for mapping floating plumes, since organics have a lower conductance than most groundwaters, and for determining the depth of contamination in general.[43,52,55] A sufficiently high contrast in conductivity between contaminated and native groundwater is necessary to obtain accurate results.[54] Resistivity works best when the surrounding soils are highly conductive,[43] and geologically uniform.[54,56] Although both EM and resistivity surveys measure related subsurface characteristics, each has unique advantages in certain situations and can be utilized accordingly by experienced personnel.

REFERENCES

1. Simpson, L.C., States set soil analysis regs, *Poll. Eng.*, 22(8), 93, 1990.
2. American Petroleum Institute, A Guide to the Assessment and Remediation of Underground Petroleum Releases, Pub. No. 1628, API, Washington, D.C., 1989.
3. California State Water Resources Control Board, Leaking Underground Fuel Tank Field Manual. Sacramento, CA, 1989.
4. Mason, B.J., Preparation of Soil Sampling Protocol: Techniques and Strategies, EPA-600/4-83-020, U.S. EPA Environmental Research Center, Las Vegas, NV, 1983.
5. Keith, L.H., Environmental sampling: a summary, *Environ. Sci. Tech.*, 24(5), 610, 1990.
6. Rajagopal, R. and Williams, L.R., Economics of sample compositing as a screening tool in ground water quality monitoring, *Ground Water Monitoring Rev.*, 9(1), 186, 1989.
7. American Petroleum Institute, Literature Survey: Hydrocarbon Solubilities and Attenuation Mechanisms, Pub. No. 4414, API, Washington, D.C., 1985.
8. Fitzgerald, J., On-site analytical screening of gasoline contaminated media using a jar headspace procedure, in *Petroleum Contaminated Soils*, Vol. 2, Calabrese, E.J. and Kostecki, P.T. Eds., Lewis Publishers, Chelsea, MI, 1989, 119.
9. Anon., Chevron unveils cleaner-burning gasoline, diesel, *Nat. Petrol. News*, 82(9), 30, 1990.
10. Potter, T.L., Analysis of petroleum-contaminated soil: an overview, in *Petroleum Contaminated Soils*, Vol. 2, Calabrese, E.J. and Kostecki, P.T., Eds., Lewis Publishers, Chelsea, MI, 1989, 97.
11. Burris, J. and Jung, P., Installation Restoration Program Analytical Protocols U.S. Air Force Occupational and Environmental Health Lab., Brooks Air Force Base, San Antonio, TX, 1988.
12. Hanie, J., Underground storage tank releases in Arizona: causes, extent, and remediation, in *Petroleum Contaminated Soils*, Vol.2, Calabrese, E.J. and Kostecki, P.T., Eds. Lewis Publishers, Chelsea, MI., 1989, 55.
13. De Angelis, D., Quantitative determination of hydrocarbons in soil, in *Manual of Sampling and Analytical Methods for Petroleum Hydrocarbons in Ground Water and Soil*, Kane, M., Ed., American Petroleum Institute Pub. No. 4449, API, Washington, D.C., 1987.
14. Bell, C.E., State-by-state summary of cleanup standards, *Soils*, Nov.–Dec., 10, 1990.
15. U.S. Environmental Protection Agency, Test Methods for Evaluating Solid Waste, SW-846, December, Washington, D.C., 1987.
16. Kostecki, P.T. et al., Regulatory policies for petroleum-contaminated soils: how states have traditionally dealt with the problem, in *Soils Contaminated by Petroleum — Environmental and Public Health Effects*, Calabrese, E.J. and Kostecki, P.T. Eds. John Wiley & Sons, New York, 1988, 415.
17. Russell, D.L. and Hart, S.W., Underground storage tanks, potential for economic disaster, *Chem. Eng.*, 94, 61, 1987.
18. Bauman, B., Current issues in management of motor fuel contaminated soils in *Petroleum Contaminated Soils*, Vol.2, Calabrese, E.J. and Kostecki, P.T., Eds. Lewis Publishers, Chelsea, MI, 1989, 31.
19. Barker, J.F. and Patrick, G.C., Natural attenuation of aromatic hydrocarbons in a shallow sand acquifer, in *Proc. NWWA/API Conf. Petrol Hydrocarbons and Organic Chem. in Groundwater — Prevention, Detection, and Restoration*, National Water Well Association, Dublin, OH, 1985, 160.
20. Chaing, C.Y. et al., Aerobic biodegradation of benzene, toluene, and xylene in a sand aquifer — data analysis and computer modeling. *Groundwater*, 27(6),823, 1989.

21. Mackay, D.L., The chemistry and modeling of soil contamination with petroleum, in *Soils Contaminated by Petroleum — Environmental and Public Health Effects,* Calabrese, E.J. and Kostecki, P.T., Eds., John Wiley & Sons, New York, 1988, 5.

22. Dragun, J. and Barkach, J., Three common misconceptions concerning the fate and cleanup of petroleum products in soil and groundwater, in *Petroleum Contaminated Soils,* Vol. 2, Calabrese, E.J. and Kostecki, P. T., Eds. Lewis Publishers, Chelsea, MI, 1989, 149.

23. Kane, M., Manual of Sampling and Analytical Methods for Petroleum Hydrocarbons in Groundwater and Soils, American Petroleum Institute Pub. No. 4449, API, Washington, D.C., 1987.

24. Foxboro Co., Laboratory Application Data and Product Specifications for the Century Organic Vapor Analyzer (OVA), 1986.

25. U.S. EPA, Field Measurements — Dependable Data When You Need It, EPA/530/UST-90-003, Office of Solid Waste and Emergency Response, Washington, D.C., 1990.

26. Buchmiller, R.C., Screening of ground water samples for volatile organic compounds using a portable gas chromatograph. *Ground Water Monitoring Rev.,* 1989, 126.

27. Robbins, G.A. et al., Use of headspace sampling techniques in the field to quantify levels of gasoline contamination in soil and groundwater, in *Proc. NWWA/API Conf. Petrol. Hydrocarbons and Organic Chem. in Groundwater — Prevention, Detection, and Restoration,* National Water Well Association, Dublin, OH, 1987, 307.

28. Holbrook, T., Vapor plume definition using ambient temperature headspace analysis, in *Proc. NWWA/API Conf. Petrol. Hydrocarbons and Organic Chem. in Groundwater — Prevention, Detection, and Restoration,* National Water Well Association, Dublin, OH, 1987, 317.

29. Griffith, J.T. et al., A new method for field analysis of soils contaminated with aromatic hydrocarbon compounds, in *Focus: Conference on Eastern Regional Groundwater Issues,* National Water Well Association, Dublin, OH, 1988, 223.

30. Robbins, G.A. et al., A field screening method for gasoline contamination using a polyethylene bag sampling system, *Ground Water Monitoring Rev.,* 9(4), 87, 1989.

31. Schlender, M.H. and Rowe, J., Soil gas survey techniques for the investigation of underground storage tanks, *Env. Prog.,* 8(4), 231, 1989.

32. Marrin, D.L. and Kerfoot, H.B., Soil gas surveying techniques, *Env. Sci. Tech.,* 22(7), 740, 1988.

33. Kerfoot, H.B., Is soil gas analysis an effective means of tracking contaminant plumes in groundwater? What are the limitations of the technology currently employed?, *Ground Water Monitoring Rev.,* 8(2), 54, 1988.

34. Spittler, T.M. et al., A new method for detection or organic vapors in the vadose zone, in *Soils Contaminated by Petroleum: Environmental and Public Health Effects,* Calabrese, E.J. and Kostecki, P.T., Eds., John Wiley & Sons, New York, 1988, 18.

35. Zdeb, T.F., Multi-depth soil gas analyses using passive and dynamic sampling techniques, in *Proc. NWWA/API Conf. Petrol. Hydrocarbons and Organic Chemicals in Groundwater — Prevention, Detection, and Restoration,* National Water Well Association, Dublin, OH, 1987, 329.

36. Tillman, N. et al., Soil gas surveys. II. Procedures, *Poll. Eng.,* 21(8), 79, 1989.

37. Tillman, N. et al., Use of soil gas surveys to enhance monitoring well placement and data interpretation, in *Proc. 3rd National Outdoor Action Conf. Aquifer Restoration, Groundwater Monitoring, and Geophysical Methods,* National Water Well Association, Dublin, OH, 1989, 239.

38. Marrin, Donn L., Soil-gas sampling and misinterpretation, *Ground Water Monitoring Rev.,* 8(2), 51, 1988.

39. Evans, O. D. and Thompson, G.M., Field and interpretation techniques for delineating subsurface petroleum hydrocarbon spills using soil gas analysis, in *Proc. NWWA/API Conf. Petrol. Hydrocarbons and Organic Chemicals in Groundwater — Prevention, Detection, and Restoration,* National Water Well Association, Dublin, OH, 1986, 444.

40. Greensfelder, R.W., Sampling soil gas in tight soils in *Proc. Conf. Petro. Hydrocarbons and Organic Chem. in Groundwater — Prevention, Detection, and Restoration*, National Water Well Association, Dublin, OH, 1989, 97.

41. Silka, L.R., Simulation of vapor transport through the unsaturated zone — interpretation of soil gas surveys, *Groundwater Monitoring Rev.*, 8(2), 115, 1988.

42. Kerfoot, H.B. et al., Measurement of carbon dioxide in soil gases for indication of subsurface hydrocarbon examination, *Groundwater Monitoring Rev.*, 8(2), 67, 1988.

43. Bensum, R.C. et al., Geophysical techniques for sensing buried wastes and waste migration: an applications review, in *Proc. NWWA/EPA Conf. Surface and Borehole Geophysical Methods in Groundwater Investigation*, National Water Well Association, Dublin, OH, 1984, 533.

44. Saunders, W.R. and Germeroth, R.M., Electromagnetic Measurements for subsurface hydrocarbon investigations, in *Proc. NWWA/API Conf. Petrol. Hydrocarbons and Organic Chemicals in Groundwater — Prevention, Detection, and Restoration*, National Water Well Association, Dublin, Ohio, 1985, 310.

45. Roberts, R.L. et al., A multi-technique geophysical approach to landfill investigation, in *Proc. 3rd National Outdoor Action Conf. Aquifer Restoration, Groundwater Monitoring, and Geophysical Techniques*, National Water Well Association, Dublin, OH, 1989, 797.

46. Gilmer, T.H. and Helbling, M.P., Geophysical investigation of a hazardous waste site in Massachusetts in *Proc. NWWA/EPA Conf. Surface and Borehole Geophysical Methods*, National Water Well Association, Dublin, OH, 1984, 618.

47. Hitchcock, A.S. and Harmon, H.D., Jr., Application of geophysical techniques as a site screening procedure at hazardous waste sites, in *Proc. 3rd Nat. Symsp. Aquifer Restoration and Groundwater Monitoring*, Nielson, D.M., Ed., National Water Well Association, Dublin, OH, 1983, 307.

48. Wagner, K. et al. *Drum Handling Manual for Hazardous Waste Sites*, Noyes Data Corp., Park Ridge, NJ, 1987, chap. 4.

49. Bellandi, R., *Hazardous Waste Site Remediation — The Engineer's Perspective*, Van Nostrand Reinhold, New York, 1988.

50. Dawson, G.W. and Mercer, B.W., *Hazardous Waste Management*, John Wiley & Sons, New York, 1986.

51. Gilkeson, R.H. and Cartwright, K., The application of geophysical techniques as a site screening procedure at hazardous waste sites, in *Proc. 2nd Nat. Symp. Aquifer Restoration and Groundwater Monitoring*, Nielson, D.M., Ed., National Water Well Association, Dublin, OH, 1982, 169.

52. Saunders, W. R. and Stanford, J.A., Integration of individual geophysical techniques as a means to characterize an abandoned hazardous waste site, in *Proc. NWWA/EPA Conf. Surface and Borehole Geophysical Methods*, National Water Well Association, Dublin, OH, 1984, 584.

53. Walsh, D.C., Surface geophysical exploration for buried drums in urban environments: application, in New York City, in *Proc. 3rd Nat. Outdoor Action Conf. Aquifer Restoration, Groundwater Monitoring, and Geophysical Techniques*, National Water Well Association, Dublin, OH, 1989, 935.

54. Rodrigues, E.B., A critical evaluation of the use of geophysics in groundwater contamination studies in Ontario, in *Proc. NWWA/EPA Conf. Surface and Borehole Geophysical Methods.*, National Water Well Association, Dublin, OH, 1984, 603.

55. Noel, M.R. et al., The use of contemporary geophysical techniques to aid design of cost effective monitoring well networks and data analysis, in *Proc. 2nd Nat. Symp. Aquifer Restoration and Groundwater Monitoring*, National Water Well Association, Dublin, OH, 1982, 163.

56. Bruehl, D.H., Use of geophysical techniques to delineate groundwater contamination, in *Proc. 3rd Nat. Symp. on Aquifer Restoration and Groundwater Monitoring*, National Water Works Association, Dublin, Ohio, 1983, 295.

CHAPTER 5

Contracting UST Disposal Services

Janet E. Robinson

CONTENTS

0-87371-402-4/93/$0.00+$.50
© 1993 by Lewis Publishers

Contracting UST Disposal Services

5.1 CONTRACTING OPTIONS

Because of the highly explosive nature of the gasoline fumes that are common in and around empty petroleum storage tanks, tank cleaning and removal should be performed only by specialized firms familiar with standard industry practices and safety procedures. As described below, some firms do the whole job of tank closure while others do only part; the tank owner must decide how he or she wishes to contract and coordinate the parties involved.

Tank closure can be divided into several general steps:

- Excavation, removal, or filling
- Tank entry and cleaning
- Site assessment (soil sampling and laboratory analysis)
- Tank transportation and disposal, if removed from site
- Waste disposal, possibly of hazardous waste

Each of these tasks requires specialized skills and/or equipment. Some companies, such as large tank installers, may have both the equipment and personnel to do all of them, while others, such as hazardous waste consultants, may subcontract the excavation and other tasks to a local firm, or work in conjunction with them if all contracting is done directly by the tank owner.

Firms providing tank closure services fall into three general categories, depending on their primary business orientation. Tank installers, the first category, will often provide removal and waste disposal services in addition to tank installation; these companies are the obvious choice when a new vessel is desired in the same location as the old. Tank installers may also provide for tank cleaning and disposal, or may rely on a tank cleaning company to remove and dispose of sludge. Many general excavation contractors who do not specialize in tank installation provide tank removal services as well.

Hazardous waste consulting firms, a second category, may perform tank closures in addition to their normal array of hazardous waste management services. These firms usually perform soil sampling and general site management, while subcontracting the excavation and hauling. Some remediation firms will perform the tank cleaning and arrange for waste disposal as well. As specialists in site assessment, hazardous waste consultants may be able to provide the advanced assessment or remediation skills necessary in the event contamination is found around the tank.

Often, however, large treatment, storage, and disposal (TSD) facilities such as landfills and solvent reclamation plants will offer full, "turnkey" services to clients who wish to use their disposal technology. These operations will supply or arrange for equipment to excavate and transport a used tank to their site, where it will then be disposed of according to their particular technique. Tank owners considering the use of a particular landfill or other facility may wish to find out if this turnkey service is available before contracting with other disposal companies who may ultimately use the same TSD anyway.

In most cases, the best procedure is to hire a single firm to perform or manage the entire operation. This strategy reduces the amount of involvement required by the tank owner, and increases the likelihood that the task will be completed on time, without coordination problems.[1]

However, tank owners familiar with tank closure issues and procedures, who are able to commit some quantity of their own time, may save some money by contracting each party directly, a strategy called "multiple prime contracting". This approach saves the 10 to 15% overhead fees a prime contractor normally adds to the subcontractor charges, but requires the tank owner to spend his or her own time arranging and coordinating service. One common scenario is for the tank owner to contract two different firms: a general contractor or tank installer to remove and dispose of the tank (or fill it), who in turn either performs or subcontracts tank cleaning, as well as a hazardous waste consultant to complete the site assessment and arrange for laboratory analysis. Good communication and explicit delegation of responsibility among the various firms, especially in regard to site health and safety, is essential to the success of this approach.

5.2 OBTAINING SERVICE

Problems in recent years with unscrupulous businesses making profits from substandard treatment of hazardous or industrial wastes make it advisable for tank owners requiring service to proceed carefully when investigating and selecting tank closure contractors. Because financial liability for environmental damages remains with the tank owner, even if the contracted firm is legitimate, preliminary investigation may help to avoid costly litigation later on.

In addition, fuel tank removal is dangerous work. Since the potential is high for explosions, fires, pit cave-ins, exposure to harmful fumes, and other health-threatening events, removal contractors must thoroughly know their jobs and the risks involved. While the removal contractors are liable for injuries to their own employees, they are not necessarily responsible for injuries to passers-by, to the tank owner's own employees, or to other nearby people; thus, the tank owner may be in legal jeopardy in the event of an accident.[2]

The means for obtaining tank closure services can differ according to the size of the job and the tank owner's resources, both in time and money. For a single job required quickly, the nearest or most available contractor may be selected out of necessity, without reference to the capabilities or cost of other contractors. For larger efforts, a formal bid package is usually prepared and distributed to several contractors, whose responses are compared according to various criteria. Procedures for obtaining services by either method are discussed in the following sections.

The first step in obtaining proposals for tank closure is to identify contractors capable of doing the work. Companies who service underground storage tanks (USTs) are listed in the telephone book under "Tanks" or "Waste Management". Directories for larger cities are useful since businesses will often travel to regions in which they do not advertise. A short interview with a company representative is usually sufficient to determine whether their capabilities include tank closure.

5.2.1 Competitive Bidding

Competitive bidding requires soliciting proposals from a number of contractors and comparing their services to determine which would be the best for the job at hand. This method consumes a larger amount of tank owner time (and hence, money) than direct selection, but ensures that the total cost is representative of the industry and allows for qualitative comparison of competing firms. For large jobs especially, the benefits usually outweigh the initial investment of time.

Competitive bidding consists basically of three steps: identification of qualified firms as discussed above, preparation and distribution of bid packages to those firms, and receipt and evaluation of proposal packages. Other steps, such as a pre-bid meeting with contractors, a site tour or contractor interviews may be added as needed.

The bid package, the heart of the competitive bidding process, must be as comprehensive and carefully prepared as possible. A detailed bid package will reduce the amount of additional information requested by individual contractors (the provision of which can bias the bidding process, unless an addendum of all questions and answers is published and distributed), and will allow contractors to estimate costs as accurately as possible. The bid package should be especially explicit in the following areas:[3]

1. **Scope** — The bid package should detail, as specifically as possible, all tasks and site information pertinent to the removal effort. This information might include the following:

   - Tank number, size, material of construction, age, and number and location of connecting pipes
   - Current and historical product storage, and estimated quantities of residual product or sludge
   - Proximity of buildings, roads, and overhead lines or awnings, and traffic patterns (include a site plan, if available)
   - Depth to groundwater (wet hole or dry hole conditions), soil conditions, location of surface water or other sensitive environmental areas
   - Special site needs (will the facility remain open during construction?) and other pertinent information

Contractors often like to visit the site before giving final bids; a single time for a general site tour may be arranged to accomplish this.

2. **Responsibilities** — The division of labor between the contractor and the tank owner must be explicitly defined, since the omission can lead to delays or more serious consequences once the job is underway. Tasks to be addressed include the following:

- Notifying utilities with underground lines of the plan to excavate (i.e., "clearing" the site for excavation)
- Notifying the appropriate authorities and acquiring any necessary permits
- Transporting and disposing of the tank and scrap piping
- Transporting and disposing of tank waste
- Filling out and signing manifests
- Supply of drum, rolloffs, or other containers for waste disposal
- Site safety (the contractor is usually responsible for this)
- General "trash" disposal: used plastic sheeting, etc.
- Barricading and traffic control
- Backfilling and repaving
- Extra excavation and soil disposal in the event contaminated soil is discovered

3. **Standards of Performance** — Federal regulations incorporate, by reference, several industry standards for tank closure. Other Occupational Safety and Health Administration (OSHA) standards apply to general construction work, to work in confined spaces, and to hazardous waste operations. These standards should either be referenced or appended to the bid package as required standards for the contractor to follow. The following standards cover closure activities:

- API (American Petroleum Institute) 1604: "Removal and Disposal of Used Underground Petroleum Storage Tanks" (2nd ed.; 1987)
- API 2015: "Cleaning Petroleum Storage Tanks"
- API 2015A: "Guide for Controlling the Lead Hazard Associated with Tank Entry and Cleaning"
- NFPA (National Fire Protection Association) 30: "Flammable and Combustible Liquids Code"
- NFPA 329: "Underground Leakage of Flammable and Combustible Liquids" (1987)
- NIOSH (National Institute for Occupational Safety and Health): "Criteria for a Recommended Standard: Working in a Confined Space"

Federal Regulations on tank closure are contained in 40 CFR 280.70. Sources for these and other standards are listed in Appendix 3.

4. **Additional Costs** — Since the closure of USTs involves many unknowns, a means should be provided for determining the general cost of additional tasks. Unit- pricing job components such as contaminated soil removal, waste disposal, additional backhoe time, or other services can help estimate costs as the job progresses.

5. **Schedule Requirements** — Many jobs have to be completed within a specified time frame because of closure deadlines, property sale, contamination problems, or other situations. Schedule imperatives should be communicated clearly in the proposal document.

From the standpoint of the tank owner, the bid package must be designed to elicit from the contractors all the information necessary to make an intelligent decision from among them. The information from the contractors, which becomes the basis for evaluation, should include the following.[3,4]

- Amount of experience in underground storage tank removal (VERY IMPORTANT)
- Certification or training commensurate with State requirements (Wisconsin, for example, explicitly defines who is qualified to collect soil samples)[4]
- Familiarity with federal, state, and local regulations
- Knowledge and practice of tank closure procedures that are consistent with industry standards
- 40-hour health and safety training for all site workers, as required by OSHA 1910.120. This applies to tank removers and consultants completing the site assessment, but not to tank installers (see Chapter 6).
- Training and experience of personnel, specifically in the areas of respiratory protection, flammable materials handling, hazardous materials handling, Department of Transportation (DOT) regulations and placarding, manifest preparation and review, and confined space entry
- Ability of the tank owner to provide a "certificate of destruction" of the tank and scrap piping
- Equipment to be used, and whether they own it or will subcontract or rent it
- The proposed plan and procedures (with reference to industry standards) of all phases of the job (tank removal/filling, cleaning, waste disposal), including the "final fate" of the tank and all other liquid or solid wastes
- Names and services of subcontractors, including laboratories, transporters, and waste treatment or disposal facilities
- Number of years the company has been in business
- Insurance coverage (insurance company's name, policy holder's name, insurance certificate holder's name, monetary limits for worker's compensation and bodily injury/property damage)
- Itemized and total cost of the project
- Assumptions and unincluded costs (e.g., additional soil removal, soil transport and disposal)
- Unit costs for additional tasks (excavation, etc.)
- Estimated schedule of completion
- References from similar, recent jobs, indicating if the jobs were completed on-time and on-budget

Proposals should ideally be issued 3 or 4 weeks or more in advance of the job. This allows contractors to mobilize and to assign the best equipment and personnel for the job, and ensures that the tank owner does not lose the potential services of a quality contractor because the contractor has prior commitments that cannot be rearranged.

Health and safety procedures are of critical importance when working with volatile materials, yet are often treated in a cavalier manner by general contractors. By law, contractors must comply with all applicable health and safety codes, and are responsible for ensuring that all personnel on the site comply with the requirements. Tank owners should insist that a single individual from the contractor's on-site crew be designated as the site health and safety officer, to simplify the chain of responsibility for safety matters (see Chapter 6).

A prebid meeting is useful on jobs where requirements or site conditions are complex. Attended, ideally, by all the bidding contractors, the prebid meeting allows contractors a chance to ask questions in a forum where the answers, which constitute additional information, are available to all. A site tour enables contractors to inspect visually the layout of the property and more accurately determine which equipment is most appropriate and what problems may be encountered. Both the prebid meeting and the site tour translate into contractor proposals with increased accuracy of both time and cost.

Contractor interviews are useful to evaluate first-hand the contractor's personnel and to pursue questions arising from the proposal. Technical approach, specific experience, and health and safety concerns can be discussed at length, providing a better sense of contractor capabilities than is possible through a written proposal only. In addition, the face-to-face interview helps the tank owner sense whether the contractor is someone with whom he or she can work easily; the interaction of differing personality types can sometimes make the difference between a efficient job and one characterized by tension and friction.

5.2.2 Direct Selection

With direct selection of contractors, the tank owner chooses a contractor without going through a formal bid process. Final costs are negotiated, rather than being subject to comparison with those of other firms. Although less ritualized than competitive bidding, direct selection nonetheless requires a transfer of the same information: the contractor needs to know all the particulars of the job, and the tank owner needs to know the qualifications of the contractor. Regardless of how highly recommended the contractor is, the tank owner should take the time to investigate the same topics as would be solicited in a proposal, to ensure that his or her interests will be served as well as possible. Once again, the legal liability associated with UST work argues for taking the extra steps to ensure that the job is performed competently and completely. Direct selection is often used for small, quick jobs where time or money does not permit competitive bidding. Direct selection is usually not allowed for jobs in the public sector.[3]

5.2.3 Contracts

Contracts can be drawn up by either the contractor or the tank owner. Typical construction documents consist of the Plans and Specifications, which describe in detail the tasks to be performed and the design of the final product, and the more general Terms and Conditions, which describe the general rules under which the contract will be executed. The Conditions may include standard sections on "Contractor Responsibilities", "Tests and Inspections", "Rejecting Defective Work", "Unforeseen Physical Conditions", "Stopping Work", and "Safety". All the essential specifications and conditions described in the proposal must be included in the contract, since the contract is the legal document with which the contractor is obliged to comply.

Most environmental contractors will work only under contracts that include a clause limiting their financial liability to a predetermined, relatively low amount, a fact which often causes great consternation to tank owners. The reason for the limitation of liability is that environmental work causes contractors to incur relatively large legal risks on behalf of the tank owner, risks for which contractors themselves cannot obtain insurance because of historical contractions in the pollution insurance industry. Moreover, because of the frequently large amounts involved in environmental lawsuits, the contractor cannot charge enough in fees to cover this risk. Therefore, most contractors will only accept a strictly-defined amount of financial responsibility for the consequences of their work, with ultimate responsibility and legal liabilities for the job residing with the tank owner.

5.2.4 Program Management

Even though a single contractor may be designated as the agent responsible for effecting tank closure, the tank owner must still take an active role to ensure that the job goes smoothly. A few tips, described below, will help the process.

Good communication is the single most important element of project management. Frequent dialogue between the tank owner and contractor will reduce the chance of misunderstandings and wasted effort. Understanding or knowledge in undefined situations should not be assumed; all details should be spelled out.

From the initial release of bid packages to the final shovelful of dirt, tank owners and contractors alike should keep thorough written documentation of all discussions and agreements. Changes to the scope of work should *always* be in writing, rather than conveyed in verbal agreements or with handshake deals, and should be confirmed by the signatures of both parties.

A photo log of the work is an invaluable form of documentation both for the regulators and for future work on the site. Since contractors can easily forget this responsibility when caught up in the demands of the job, photographing is best delegated to a member of the tank owner's staff or to contractor personnel in a supervisory role.

The contractor is generally responsible for initiating, maintaining, and supervising safety programs and precautions at the site.[3] Safety considerations apply to everybody at the site, tank owners included; tank owners should therefore be prepared to comply with the contractor's site health and safety requirements when inspecting work in progress.

Remembering the hierarchy of contractual relationships is important when subcontractors are involved. Technically, subcontractors have a contractual relationship only with the prime contractor, not with the tank owner. Thus, unless arrangements are made otherwise, the tank owner technically should communicate to the subcontractors through the prime contractor.

Finally, at the close of the project, the tank owner should receive all records, photographs, receipts, manifest copies, laboratory results, sample location logs, and other pertinent information at the close of the project. In the event of an inquiry by the regulators or prospective property buyers, the tank owner, not the contractor, will need the information and must be able to document fully that the job was satisfactorily performed.

REFERENCES

1. University of Wisconsin, Center for Professional Development, Underground Storage Tank Installation Short Course. Course Handbook, University of Wisconsin, Madison, Wisconsin, 1989, chap. 2.
2. Anon., Tank removal: a growing hazard, *Natl. Petrol. News,* 82(9), 32, 1990.
3. Thompson's Underground Storage Tank Guide, Thompson's Publishing Co. (continuously updated; initially pub. 1987).
4. State of Wisconsin, Dept. of Industry, Labor, and Human Relations, Tank owner's guide for underground storage tanks, Madison, Wisconsin, 1990.

CHAPTER 6

Health and Safety

Janet E. Robinson

CONTENTS

0-87371-402-4/93/$0.00+$.50
© 1993 by Lewis Publishers

Health and Safety

6.1 GENERAL GUIDELINES

Health and safety concerns during underground storage tank (UST) removal and remediation are numerous, significant, and frequently disregarded by contractors with all ranges of experience. A number of deaths have resulted from explosions of gasoline tanks during removal, and the fact that tank removal is usually perceived as a routine activity, and was formerly performed by general contractors with no particular training, helps obscure the real dangers involved with the work, especially if flammable materials are involved. For this reason, the tank owner needs to be aware of precautions that both he or she, and the contractor, must observe to ensure that the removal is completed safely.

Regulations promulgated by the Occupational Safety and Health Administration (OSHA) on March 6, 1989 (29 CFR 1910.120) specify that contractors whose employees are exposed to hazardous waste and hazardous substances must have a comprehensive health and safety program consisting of a written health and safety plan, extensive training and medical monitoring of site workers, and other components designed to minimize the hazards of working with and around hazardous chemicals. The requirements of these regulations must be followed by the tank remover, tank cleaners, and other contractors whose employees may be exposed to hazardous waste or hazardous substances. More specific information about the OSHA standard is provided later in this chapter.

6.2 SITE HAZARDS

Hazards associated with tank removal can be grouped into five categories:

- Fire and explosion
- Hazardous fumes and materials
- Confined spaces
- Open holes
- Heavy equipment

Concerns and precautions in regard to each of these are presented below.

6.2.1 Fire and Explosion

During the removal of gasoline tanks and any other tanks containing highly flammable solvents or other "Class A" products, the dangers from fire and explosion rank as one of the greatest concerns. A brief discussion of the components of combustion will help illustrate why and when this is the case.

Fire requires three components to exist: fuel, oxygen, and an ignition source. In the case of flammable products, the initial "fuel" is the vapor within the tank, present as the result of residual product or sludge. In order to prevent explosion, vapor concentration levels must be either below the stored compound's "Lower Explosive Level" (LEL), below which the vapor:air mixture is too lean to propagate flame, or above a specific upper limit, where the mixture is too rich. LELs vary by compound, but the goal of vapor-freeing is to reduce the concentration of ignitable fumes, and often oxygen as well, to a level too low to support ignition. Until this is done, the tank potentially needs only a stray spark or flame to explode with deadly force. The hazard of tank explosion and fire exists up until the time that the tank is vapor-freed, and even afterwards if the tank is allowed to sit before cleaning and resaturate the internal spaces with fumes. Large voids in contaminated soil may also become rich in flammable vapors, which can be ignited by a stray match or cigarette. During vapor-freeing, especially by water-filling or eductor fans, flammable vapors forced out from the tank may accumulate, causing flammable conditions around the vent pipe or other openings. Finally, piping, drums, hoses, or other equipment used to hold or transport flammable liquid are all a potential source of flammable fumes.

However, the use of standard safety procedures will minimize the risk from fires and explosions. Some of these are listed below.

1. NO SMOKING should be allowed by anyone on the site.
2. All equipment liable to be the source of a spark from built-up static electricity or from other sources should be bonded and grounded before operation, including pumps, fans, cleaning hoses, and similar equipment used around the tank. This equipment should be located upwind from the tank.
3. The tank interior should be monitored with a combustible gas indicator (CGI) or other instrument to determine precisely the vapor concentration within the tank.
4. The tank should be removed from the excavation by lifting it straight out of the hole; flammable liquid tanks should never be dragged or pushed over the ground.
5. The tank should never be struck with a metal object to remove clinging dirt or for other purposes.
6. The air around the site should be monitored, especially if fumes from vapor-freeing are ejected close to ground level.
7. Even after the tank has been vapor-freed, only nonsparking cold cut methods should be used to cut through tank walls.

6.2.2 Hazardous Fumes and Materials

The vast majority of USTs nationwide are used to store petroleum products and other chemicals that, though commonplace, can nonetheless cause significant detrimental health effects if certain exposure levels are exceeded. Routes of exposure include dermal (through the skin), inhalation, and ingestion, the latter from contaminated dirt, dust, or residues present on food consumed on the jobsite. With prolonged exposure, the kidneys, liver, and central nervous system may be affected or the chance of developing cancer may increase; for instance, benzene, a component of gasoline, is a known carcinogen. Likewise, organic lead from leaded gasoline can cause brain damage with prolonged exposure. Symptoms vary with the specific compound, but often include headache, nausea, drowsiness, and blurred vision; loss of consciousness and death may follow after extreme or prolonged exposure. For these reasons, any chemical, regardless of how frequently used, should be treated with caution, and every effort made to prevent exposure of workers or other site personnel. As a general procedure, personnel should avoid handling chemicals without protective equipment: chemical resistant gloves, respirators if necessary, and chemical resistant suits. OSHA regulations for hazardous waste work require such precautions. Workers should know the characteristics of the chemicals they must deal with; this may require that the tank owner provide the removal contractor with the Material Safety Data Sheets (MSDS) for the compounds currently or formerly stored in the tank.

No eating or drinking should be allowed within 50 ft of the worksite, and employees should be required to wash their hands before meals or snacks. Open containers of food, such as coffee and the ubiquitous box of doughnuts, should not be left around the site.

When ground-level fumes are possible, air monitoring should be conducted to ensure that levels do not exceed safe concentrations. One point where this may occur is while the tank is being pumped free of product: vapor concentrations may become quite high around the tank access hole and the receiving drums, requiring that individuals in the area don respirators, with appropriate cartridges, before continuing work. The safety officer is responsible for monitoring fumes and, when necessary, requiring the use of respirators or other protective measures.

6.2.3 Confined Spaces

Broadly speaking, a confined space is any area with physical features that cause vapors to accumulate to dangerous levels. On a tank removal job, the tank itself is the most obvious confined space: because its internal space may be both high in fumes and low in oxygen, workers generally enter this space with an independent air supply, normally from compressed air containers outside the tank. The tank excavation itself may present similar problems; fumes emanating from contaminated soil may reach high levels at the bottom of the hole, while fumes that are heavier than air will also settle there and displace the oxygen. Heavier-than-air fumes can in fact settle in any low area around the site if natural wind to disperse them is absent; extra precautions are therefore necessary to prevent their accumulation around the site.

Confined space entries are most often required during tank cleaning, and since personnel must be specially trained, the tank owner is normally not involved in this procedure. Standard procedure involves the use of the buddy system during entry: the individual inside the tank is assisted by another worker outside the tank who is dressed to the same level of personal protection and who can enter the tank immediately if help is needed. Personnel inside the tank are attached to lifelines by which they can, if necessary, be hauled out of the tank. Workers may carry their air supply on their back or attached to their belt, or be connected by an airline to a cascade air supply system outside the tank.

Guidelines for working in confined spaces have been published both by the National Institute for Occupational Safety and Health (NIOSH) and by the American National Standards Institute.

OSHA is currently working on formal standards for confined space entry; these standards are due for publication shortly.

6.2.4 Open Trenches

At some point during tank removal, the tank owner will be faced with one or more large, open, and possibly deep holes on his or her property. If laboratory analyses are necessary, these holes may be open for days or weeks, presenting a safety hazard to personnel and nearby traffic. Over time, the sides of open holes frequently cave in, undercutting asphalt surfaces which may then break if someone inadvertently walks over them. Site workers or neighborhood children can fall into deep holes and be unable to climb out, with the very act of trying to do so precipitating further side-slippage. The danger to personnel of burial in a deep trench or excavation is very real; sand weighs approximately 1 ton/yd^3, and unshored excavations often drop more than that in a single cave-in. Trapped individuals are usually unable to free themselves when caught under collapsing soil.

OSHA construction standards at 29 CFR 1926 Subpart P (i.e., 1926.651 and 1926.652) explicitly address entry of open trenches and pits. According to those regulations, trenches more than 5 ft deep cannot be entered without shoring the sides or using a portable trenchbox. Additional detailed regulations describe the type of shoring to be used in various circumstances; the excavator or prime contractor should be familiar with these. If samples must be collected from the bottom of a deep hole, the contractor should use a long-handled hand auger, or should ask the backhoe operator to scoop up a bucket of dirt with the backhoe. Personnel should avoid entering the pit for any reason.

The hole should be barricaded with barrier tape or sawhorses whenever work is not in progress. When the holes are next to a road, heavy equipment or other vehicles may be parked between the hole and the road to prevent a car from crashing into the hole during an accident. If the hole will remain undisturbed for a long period, the tank owner may prefer to fill it in, with the knowledge that, in the future, reexcavation may be necessary to proceed with remediation.

6.2.5 Heavy Equipment

Heavy equipment is common to most construction sites, and workers soon become accustomed to the noise and normal backup alarms associated with them. This familiarity itself can present a hazard; a worker engaged in conversation may step too near the vehicle tracks or be struck by a swinging bucket. Additionally, the loud noise produced by large equipment not only presents a threat of ear damage, but may also drown out warning shouts or calls for help. Other power tools, from jackhammers to drills and circular saws, present their own range of dangers, from flying debris to sharp edges.

The best preventative is constant diligence. Workers should stand clear of heavy equipment, and move away if something else, such as conversation or paperwork, requires their attention. No other activities should be conducted in the area of operating equipment; equipment operators focusing on their work cannot closely track the location of other personnel and vehicles.

Ear protection should be worn by individuals exposed to high levels of noise. Good communication, either by voice, eye contact, or hand signals, should be maintained between all members of the crew. All members should additionally understand fully the nature and sequence of tasks completed at the site. Safety equipment, such as hardhats and steel-toed boots, should be worn at all times.

Finally, the number of personnel on site should be minimized, especially during tank purging when explosive fumes are likely to accumulate on the site. Spectators should be kept at a distance, and the tank owner and station employees, unless conferring with the site workers or inspecting work, should also remain clear. Family members and other employees should be discouraged from standing around the work area unless their services are needed. A removal contractor's insurance will usually cover injuries to his or her own crew, but compensation for injuries to others may involve lengthy lawsuits.

6.3 29 CFR 1910: OSHA HEALTH AND SAFETY REQUIREMENTS

On March 6, 1989, OSHA finalized its regulations for hazardous waste operations and emergency response. These regulations apply to workers involved in the following activities (among others), as summarized from 29 CFR 1910.120(a):

1. Cleanup operations, required by a government body, involving hazardous substances that are conducted at an uncontrolled hazardous waste site, including initial investigations at government-identified sites which are conducted before the presence or absence of hazardous substances has been ascertained
2. Corrective actions involving cleanups at RCRA sites
3. Voluntary cleanups at uncontrolled hazardous waste sites
4. Operations involving hazardous wastes that are conducted at treatment, storage, and disposal (TSD) facilities
5. Emergency response operations

"Hazardous substance", as defined in the regulations, includes substances defined as hazardous by both the Comprehensive Environmental Response, Compensation, and Liability Act (CERCLA) and the Department of Transportation, which in turn includes petroleum. In addition, many states consider used oil to be a hazardous waste. For these reasons, the regulations apply to removers of petroleum tanks, since petroleum is a hazardous substance and is often present in the soil around the tank. However, tank installers who are not involved with waste product or contaminated soil would not be required to comply with these OSHA regulations.

Tank owners should be sure that the contractors they hire to remove the tank are in compliance with these regulations, unless the tank did not hold hazardous substances or the tank owner is certain that the contractor will not be exposed to hazardous substances, either residual in the tank or released around it. Since the latter assurance can rarely be made, the OSHA regulations at 29 CFR 1910.120 should be a requirement of all prospective contractors.

If a single prime contractor, such as a tank installer or general contractor, is hired who subsequently hires a subcontractor to remove the tanks, the prime contractor must ensure that the tank remover has a health and safety program in compliance with 29 CFR 1910.120. Because the prime contractor is liable for all work performed by the subcontractor, the prime contractor must therefore ensure that the tank removal subcontractor is complying with its own health and safety plan. If the subcontractor is not complying with its own plan, then the prime contractor may need to step in to enforce the health and safety program simply to protect his or her own interests.

Some of the general health and safety requirements of 29 CFR 1910.120 that are of importance to UST removers are described below. For a more complete account, the regulations should be consulted directly.

6.3.1 Health and Safety Plan

UST removers must have both a general and a site-specific health and safety plan. The general plan should include copies or descriptions of the following:

- An organizational structure
- A comprehensive workplan
- A site-specific health and safety plan (which need not repeat the employer's standard operating procedures noted below)
- The health and safety training program
- The standard operating procedures for health and safety
- Any necessary interface between general program and site-specific activities

The site-specific health and safety plan contained in the general program described above should be maintained on the workplace and should address all the potential health and safety issues at that particular site. As noted in 29 CFR 1910.120(b)(4), the required elements of the site-specific plan are as follows:

- A hazard analysis for each task
- Employee training assignments
- Required personal protective equipment
- Medical surveillance requirements
- Air or environmental monitoring techniques
- Site control measures
- Decontamination measures for personnel and equipment
- An emergency response plan
- Confined-space entry procedures
- A spill containment program

6.3.2 Site Characterization and Analysis

In order to identify hazards and evaluate the need for personal protective equipment, the tank remover should conduct a preliminary evaluation of a site's characteristics prior to initiating work.

6.3.3 Training

The need for training is at the heart of the OSHA regulations and is probably their best-known component, since obtaining training for all site workers incurs considerable expense for the contractor. Training requirements are as follows:

For routine site employees	40 hours initial training
	24 hours field training
	8-hour annual refresher
For routine site employees with minimal exposure or professionals occasionally on-site (land surveyors, geophysical surveyors, etc.)	24 hours initial training
	8 hours field training
	8-hour annual refresher

Managers in either category must have the same training, plus 8 hours of supervisor training as well.

6.3.4 Medical Surveillance

All workers exposed to hazardous substances at or above permissible exposure levels should be included in a medical surveillance program that includes an entrance physical, an annual physical tailored to the hazards of the job, and an exit physical. These records must be retained long-term by the employer.

Other components of a health and safety program as outlined by 29 CFR 1910.120 include requirements for engineering controls, environmental monitoring, personnel decontamination procedures, and a variety of other topics.

Because these requirements were formulated with regard to hazardous waste site investigations, such as those that take place at Superfund sites, the requirements may seem excessive for a "simple" tank removal; however, as discussed earlier, the chemical and physical dangers involved in tank removal work are significant. These regulations require contractors to formally quantify and respond to these hazards, and are now routinely followed by all conscientious tank removal and cleaning firms.

The specific safety procedures to be followed during every stage of a tank removal job are too numerous to list here, but additional emphasis on the importance of insisting on compliance with the site health and safety plan is a valuable final message. If a tank owner observes a practice which he or she believes to be unsafe, the job supervisor or site Health and Safety Officer should be approached immediately. If the situation persists, the tank owner can order the work stopped until consultation with other contractor personnel (e.g., management) or until a call to OSHA resolves the issue. In addition to the obvious personal tragedy of an injury or death on site, a construction accident will delay the job completion and may result in lengthy lawsuits and counter-lawsuits later on. Insisting on a continual observance of preventive safety practices, even when they slow the work and elicit grumbling from the workers, is always the most prudent course of action.

PART II
UNDERGROUND STORAGE TANK
FINANCIAL ASSURANCE

CHAPTER 7

UST Financial Responsibility: A Fact of Life in the 1990s

CONTENTS

0-87371-402-4/93/$0.00+$.50
© 1993 by Lewis Publishers

UST Financial Responsibility: A Fact of Life in the 1990s

7.1 INTRODUCTION

Owners and operators of underground storage tanks (USTs) within the retail motor fuel industry faced two important problems in 1990 that could prove to change forever the financial profile of that industry. The possible impact of these same two problems on UST-owning firms outside of this industry — farms, local governments, and many businesses with vehicle fleets — could be equally severe.

First, the Persian Gulf crisis quickly led to steep increases in the price of crude oil, resulting in lower profit margins for many small retailers and the potential for a prolonged decrease in consumer demand. Second, two federal deadlines came and went that would have required firms with between 1 and 12 and 13 and 99 USTs to demonstrate $1 million of financial responsibility for cleanup and third-party liability costs associated with tank failure.[1] Although crude oil returned to its pre-invasion price in early 1991, the U.S.-Iraqi War, which began January 16, 1991 and ended less than 2 months later, pointed to the increasingly unstable conditions that surround the market for crude oil. And in April 1991 another — already extended — financial responsibility compliance deadline passed for firms with 13 to 99 USTs.

Although lacking the public attention given to the war in the Middle East, many tank owners and operators were painfully aware of these deadlines, especially since they potentially could produce even greater financial burdens on the smaller and less diversified firms that dominate the total UST universe. Only last-minute extensions by EPA of the 1990 deadlines prevented many of these owners from violating the agency's financial responsibility regulations and facing fines of up to $10,000 for each day of noncompliance.[2] Now, those firms with 13 to 99 USTs without appropriate financial responsibility mechanisms are subject to enforcement actions by U.S. EPA and authorized states. In April 1991, the deadline for petroleum marketers with less than 13 UST (as well as marketers with less than 100 USTs at a single facility and nonmarketers with a net worth of less than $20 million)* was extended by U.S. EPA, although at that time the agency did not state by how much.[3]

---

* As discussed in greater detail in Chapter 8, the UST regulations generally group owners/operators engaged in petroleum marketing and with fewer than 13 USTs at multiple locations or fewer than 100 USTs at a single location with nonmarketers with a net worth of less than $20 million, regardless of the number of USTs owned or operated by the nonmarketer.

On January 30, 1991, three major associations representing medium and large UST-owning firms* asked the Administrator of U.S. EPA, William K. Reilly, to push compliance deadlines back to December 31, 1992 for all firms with fewer than 1000 tanks, citing potentially unreasonable economic impacts as justification.[4] (UST owners/operators with 100 to 999 USTs were required to comply with the financial responsibility regulations on October 26, 1989.[1])

In fact, EPA did not extend deadlines for all UST owners with fewer than 1000 tanks but the agency gave owners with fewer than 13 tanks (as well as marketers with less than 100 USTs at a single facility and nonmarketers with a net worth of less than $20 million) more time than they had hoped for. Although EPA proposed in August to extend the compliance deadline for this group of UST owners from October 26, 1991 to December 31, 1992, the agency shocked most observers by its final decision (made in December 1991) to extend the deadline to December 31, 1993.**

7.1.1 Coping with UST Financial Responsibility

Compounding the burden of the financial responsibility requirements are a host of additional deadlines that require UST systems to be upgraded to meet minimum design, construction, release detection, installation, corrective action, and closure standards.[5] While federal regulations allow these standards to be phased in well after the financial responsibility regulations, some states (and localities), who are charged with implementing the national UST program, may choose to accelerate deadlines for tank upgrading. More important, the need for uncontaminated sites and sound USTs required for coverage under the most commonly used financial responsibility mechanisms may force many firms to meet these technical requirements sooner than required under the federal program.

Congress, EPA, and the states continue to propose methods to soften the financial impact of the requirements, including reducing requisite levels of financial responsibility, extending compliance deadlines, suspending enforcement efforts, and establishing various state "assurance" funds and loan programs for tank upgrades. While most of these approaches offer some relief, and some clearly are more promising than others, demonstrating financial responsibility for UST leaks remains difficult for most firms in the regulated community. This problem persists because regardless of how current approaches differ

---

* These industry associations were the National Association of Convenience Stores, Petroleum Marketers Association of America, and Society of Independent Gasoline Marketers of America.

** See, U.S. EPA, Environmental Fact Sheet — EPA Concerned About Small Businesses: Extends Compliance Date For Underground Storage Tank Financial Responsibility, August 1991; 56 FR 40292–40295; 56 FR 66396. See also, UST Financial Assurance Deadline Extended 26 Months for Small Firms, *Groundwater Protection News,* January 7, 1992, p.5.

on the surface, they all ultimately require significant financial resources or sound USTs (or both) in order to qualify for coverage — conditions that simply are not met by many firms.

Part II of this book describes the nature of the financial responsibility dilemma facing both the public, who has both a financial and environmental stake in ensuring that leaking USTs are effectively remediated by the responsible parties, and the regulated community, who must follow whatever course of action is established by U.S. EPA and the states. Chapter 7 describes the rationale and evolution of UST financial responsibility regulations. It also describes many of the problems that have been encountered in developing a viable system to ensure that funds are available for cleaning up leaking USTs and that these funds are managed in an equitable and efficient fashion.

Chapter 8 describes the current UST financial responsibility regulations in detail. The federal Leaking Underground Storage Tank (LUST) Trust Fund and corrective action regulations for responding to UST leaks and spills also are covered. Chapter 9 focuses on the use of insurance, the most critical — and in many ways problematic — option for demonstrating financial responsibility. Eligibility requirements and insurance availability issues are important features of Chapter 9, along with a discussion of how one state, Washington, is attempting to improve the availability of LUST insurance.

Chapter 10 addresses the other most important financial responsibility option — state funds. While such funds are relatively common, they differ dramatically in the types of coverage provided. Chapter 10 discusses the advantages and disadvantages of various features of state funds and describes the approval process that U.S. EPA follows to determine if a state fund can be used to demonstrate all or a portion of an UST owner/operator's financial responsibility. Four case examples are used to illustrate some of the current variation in how states structure and administer such funds.

Finally, Chapter 11 addresses common law liability issues of leaking USTs. Although rarely discussed, one of the most important reasons for UST owners/ operators to have a financial responsibility mechanism is to avoid financial ruin in the face of a tank leak that injures or damages a third party. Even with appropriate financial responsibility, UST owners/operators have a crucial interest in protecting their assets (including their financial responsibility mechanism). Third parties affected by a leaking UST also need to know what types of claims, evidence, and damages must be argued in order to win a suit brought against an UST owner/operator. Chapter 11 discusses some of the trends in common law that have emerged in cases where civil suits have been brought against UST owners/operators because of alleged damages from LUSTs.

7.1.2 Requirements for Public Acceptance of UST Financial Responsibility

Under these circumstances, many firms will require high levels of public funding to demonstrate financial responsibility, which is not likely in a time of rising taxes, sluggish economic growth and increasing environmental awareness. In order to meet with public acceptance, any financial responsibility option most likely will have to meet two criteria. First, the majority of costs associated with compliance — including any UST upgrades needed to qualify for coverage — should be born by the regulated party (the so-called "polluter pays" principle). Second, any approach that is to succeed must provide the tank owner or operator an adequate financial incentive to maintain high quality UST systems and operating procedures.

Even those approaches that seemingly require only minimum financial commitment from — and provide only minimum financial incentives to — the tank owner or operator, such as certain types of state funds, eventually will conform to these criteria. And even if such funds bring the regulated community into compliance initially, they will be of little value to UST owners or the public if they are not politically or economically feasible over time.

Together, these two criteria will tend to make it difficult for those firms with older tanks and relatively low profits to meet financial responsibility requirements. Simply put, for most UST owners and operators, there is no "easy way out" of the problems posed by the financial-responsibility regulations.

In addition, to meet the business needs of UST-owning firms, financial responsibility mechanisms also must meet a third criteria: they must provide the financial resources needed for firms to remain in business in the event of an UST leak — an important consideration given that the federal UST program will lead to increased leak discoveries. Eventually, any acceptable requirements probably will shift the UST universe to firms with higher profits and newer, more sophisticated UST systems. Tank owners/operators participating in the development of UST financial responsibility programs at the state level should recognize that such a shift is not necessarily contrary to public opinion or the intent of the federal UST program — although it cannot be said that EPA and state regulators have not recognized the potential for severe financial hardships in the regulated community.

7.1.3 Improving the Chances for Compliance with UST Financial-Responsibility Requirements

EPA has yet to enforce its UST financial-responsibility requirements, and while many states (which are charged with implementing EPA's UST program) have established enforcement and penalty procedures, it is too soon to tell whether they will be applied stringently after final compliance deadlines. In fact, only two states — Mississippi and New Mexico — have had their UST programs approved by EPA. Thus, unless feasible, longterm compliance options are developed, the future of many existing firms using USTs may depend on the leniency of federal and state regulatory officials with the power to levy fines and other penalties. Unfortunately, officially postponing the enforcement of the UST financial-responsibility requirements, although an option, has not been viewed favorably by EPA thus far.

In the meantime, UST owners and operators should understand the issues surrounding the financial responsibility debate along with the options they have for meeting UST financial-responsibility requirements, especially the options most likely to be used by those with fewer than 100 tanks — private insurance and state assurance funds. They also should understand that the principal groups involved in shaping UST financial responsibility policy — tank owners, private insurers, bankers, and state and EPA officials — often view the problems of and solutions to financial responsibility very differently. Thus, because what one group advocates often is perceived as threatening another group's vital interests, the financial responsibility debate has taken on an important political dimension.

7.2 BACKGROUND OF UST FINANCIAL RESPONSIBILITY REQUIREMENTS

The current federal UST program evolved from two laws passed in the mid-1980s. Congress enacted the Hazardous and Solid Waste Amendments of 1984 (HSWA) to require USTs containing petroleum products and hazardous waste to meet certain technical and operating requirements necessary to protect human health and the environment.[6] HSWA also authorized EPA to establish financial responsibility regulations requiring that UST owners and operators have funds available for the corrective action and third party compensation that might result from an accidental UST spill or leak.[6]

The Superfund Amendments and Reauthorization Act of 1986 (SARA) was passed 2 years later to require EPA to establish financial responsibility regulations (it had been only a discretionary duty under HSWA) and set a minimum level of $1 million of financial responsibility per occurrence for tanks that are "engaged in petroleum production, refining, or marketing."[7] SARA also established a $500 million federal LUST Trust Fund to use for corrective action (not third party compensation) before the financial responsibility regulations were published and afterwards, under certain circumstances when the financial resources of owners or operators were inadequate despite compliance with financial responsibility requirements.[7]

7.2.1 The Environmental and Legal Context of LUST Financial Responsibility

Both laws were enacted in response to increased discoveries of LUSTs and the resulting burden placed on federal and state governments charged with ensuring a safe environment for their citizens. While the size of the LUST problem is not known precisely, it generally is thought to be a principal source of groundwater contamination. The U.S. EPA estimated in 1986 that there were 1.4 million USTs at 500,000 facilities in the U.S.[8] This estimate was increased to "several million" in an EPA document released in 1988.[9] More recent figures put the number of USTs at 2 million.[10] Approximately 96% of these tanks store petroleum products and over half of these belong to the retail motor fuel industry.[8,11] With such a huge number of potential sources of contamination, it is not surprising that public officials would be concerned with the problems — both environmental and financial — posed by leaking USTs.

7.2.1.1 The Scope of Contamination from LUSTs

As with the total number of USTs, EPA estimates of those currently leaking or in eminent danger of leaking also have varied widely — between "tens of thousands" and "hundreds of thousands."[9,11] In 1986 EPA stated that 35% of nonfarm USTs storing motor fuel would fail a "tank-tightness" test.[12] A regulatory document prepared for EPA in 1987 estimated that almost 90% of USTs containing petroleum were "unprotected, bare steel" and one third of these were thought to be over 20 years old.[8,13] Up to 200,000 of these USTs were leaking with many more expected to leak in the near future.[11]

In the mid-1980s many state environmental agencies were finding that LUSTs were the leading source of groundwater contamination. For instance, in 1985 Virginia reported that 97 of 129 groundwater pollution incidents were petroleum contamination, mostly from LUSTs.[14] And reports of groundwater contamination by states continued to rise steadily during the 1980s, partly because of increased groundwater monitoring activities.[15]

Together with the federal LUST estimates, the states' experience supported the concern that the problem was widespread. In addition, the debate over LUSTs at the state level took place within a broader context of heightened public concern over groundwater contamination and the development of state-wide groundwater quality protection plans.

7.2.1.2 Public Involvement in UST Financial Responsibility

The support for UST regulation was driven generally by public concern over the potential for widespread groundwater contamination. The specific need for financial responsibility requirements, however, was driven by state concerns that only a fraction of existing LUSTs were being detected and, even for these tanks, no stable mechanism — either judicial or regulatory — existed to ensure that responsible parties addressed the resulting contamination.

State resources for assuring cleanup of even these relatively few sites were still far from adequate. State officials reasoned that the number of leaks that were discovered almost certainly would continue to increase because EPA's upgrade regulations required release detection programs and (in many cases) eventual tank replacement. In fact, many LUSTs have been discovered due to the tank replacement or closure activities encouraged by the federal UST program.

Under a scenario of dramatically higher LUST discoveries, states faced three choices: (1) ignore a dramatic increase in reports of groundwater contamination; (2) obtain unprecedented levels of public funding devoted exclusively to cleaning up groundwater; or (3) require that UST operators themselves be financially capable of addressing contamination problems. In addition to the high expense associated with corrective action to clean up a contaminated site, the inherent risk to the public that resulted from UST operation made third party compensation a logical addition to any financial responsibility requirements (in much the same way that it does for owning an automobile). Given these choices, it is not surprising that a political consensus emerged to mandate that the cost of owning or operating USTs should include the financial capacity to address contamination from the tanks.

7.3 EVOLUTION AND SUMMARY OF EPA'S UST FINANCIAL RESPONSIBILITY PROGRAM

EPA first proposed financial responsibility regulations for USTs containing petroleum products on April 17, 1987 and final regulations were published October 26, 1988.[16,17] (EPA's final UST financial regulations are described in greater detail in Chapter 8.) The regulations applied to all USTs containing petroleum except farm and residential tanks with a capacity of 1100 gallons or less, home heating oil tanks, septic tanks, and other obvious candidates for exemption.[17] The principal groups covered by the proposed — as well as final — regulations include the retail motor fuel industry (refiners, wholesalers, independent marketing chains, and open dealers) and nonmarketers who use motor fuel and other petroleum products (including businesses maintaining vehicle fleets, other businesses using USTs for nonmarketing purposes, farms, and local governments).[17]

From the beginning, it was clear that the regulated parties with the most to lose were the UST owners and operators without large numbers of USTs or large financial reserves: small petroleum marketers and many nonmarketers, including farmers with USTs, local governments, and small commercial entities with USTs used to refuel business vehicles. Tracing the evolution of EPA's final regulations, it is apparent that most changes made to accommodate the concerns of the regulated community have not addressed all of the problems faced by those with the very fewest USTs (i.e., 1 to 12 tanks). Unlike EPA's air pollution and water pollution programs which — because of their multibillion dollar price tags for big business — have been opposed vigorously (and often successfully) by the nation's largest corporations, the federal UST financial responsibility program is a problem centered primarily in the small business community.

7.3.1 EPA's Initial Proposal — April 17, 1987

The general objective of EPA's proposal was to ensure that tank owners would have sufficient funds for the costs associated with 99% of accidental UST releases.[18] Based on the authority provided by HSWA and SARA, EPA proposed regulations establishing various classes of UST owners and operators based on the number of USTs owned and operated and set minimum amounts of financial responsibility for each. All classes were to maintain $1 million of financial responsibility *per occurrence* (available for a single accidental release) for corrective action and third-party compensation.[18] In addition, EPA proposed that each class maintain different *yearly aggregate* levels of financial responsibility. The classes and aggregate amounts of financial responsibility in EPA's original proposal for financial responsibility were:[18]

- 1 to 12 Tanks: $1 million annual aggregate
- 13 to 60 Tanks: $2 million annual aggregate
- 61 to 140 Tanks: $3 million annual aggregate
- 141 to 250 Tanks: $4 million annual aggregate
- 251 to 340 Tanks: $5 million annual aggregate
- Over 341 tanks: $6 million annual aggregate

Tank owners and operators could choose from the following mechanisms to demonstrate financial responsibility:

- Financial test of self insurance
- Guarantee
- Indemnity contract
- Insurance/risk retention group
- Surety bond
- Letter of credit
- State required mechanisms
- State assumption of responsibility

Except for the insurance/risk retention group and state assumption of responsibility options, the allowable mechanisms generally required that the UST owner or operator himself provide the full financial resources required for coverage.

As discussed in greater detail below, a number of comments from tank owners and operators, insurance industry representatives, and state and local governments convinced EPA that these and other aspects of the financial responsibility proposal should be modified. In addition, one component noticeably absent from the proposed program was a schedule for compliance with the regulations, once they were finalized. After approximately 1 1/2 years, EPA published its "final rule" setting the financial responsibility requirements for USTs containing petroleum.[17] The final requirements, while providing some relief to the regulated community, still present significant obstacles to many UST owners and operators, particularly those with older tanks and little disposable income.

7.3.2 The Final Financial Responsibility Regulations — October 26, 1988

After a vigorous lobbying from a variety of "interested parties", EPA published the final regulations for the UST financial responsibility program.[17] The changes in the regulations were designed to address three guiding objectives.

First, the program "must require adequate and reliable financial assurance for the costs of UST releases" based on the certainty that funds will be available, the sufficiency of required financial responsibility levels, and the availability of funds for financial responsibility. Second, the program is supposed to provide "flexibility, where possible, to increase the feasibility of compliance by the regulated community" while still requiring adequate and reliable financial assurance. Finally, the UST financial responsibility program should "promote expansion of existing mechanisms and development of new ones to achieve maximum compliance by UST owners and operators."[19] Unfortunately, only time will tell if the final program requirements meet these objectives.

Although the program retained the basic approach established in EPA's initial proposal, many changes were made, some with significant implications for UST owners/operators. The most important changes, however, resulted in lower levels of financial responsibility for some UST owners/operators and a schedule for compliance with the regulations.

7.3.2.1 Required Levels of Financial Responsibility

The final regulations eliminated most of the UST categories described in the previous section. Current requirements are that *all petroleum marketing facilities* must maintain $1 million of financial responsibility on a per occurrence basis while nonmarketers handling less than 10,000 gal of product per month are required to maintain $500,000 of financial responsibility.[20] Petroleum marketing facilities are defined as: "all facilities at which petroleum is produced or refined and all facilities from which petroleum is sold or transferred to other petroleum marketers or the public."[21]

The difference in per occurrence coverage between petroleum marketing and nonmarketing facilities was possible because the HSWA authorized EPA to set financial responsibility limits for nonmarketers below the $1 million minimum established by the law.[22] EPA set this lower per-occurrence limit for these low-volume nonmarketers for two reasons. First, the agency determined that a higher limit was not needed to assure that 99% of all leaks would be cleaned up adequately. Second, EPA reasoned that a lower limit would improve the likelihood that financial responsibility mechanisms would become more widely available to these UST owners/operators.[18]

In addition to changes in the amount of per occurrence coverage, EPA also reduced the amount of annual aggregate coverage required — at least for many tank owners. The new annual aggregate levels of financial responsibility required are

- 1 to 100 tanks: $1 million annual aggregate
- 101 or more tanks: $2 million aggregate.

The requirement for annual aggregate coverage does not distinguish between marketing and nonmarketing facilities.

The reasons for the changes made in required aggregate financial responsibility were based on the availability of financial mechanisms and the economic impacts on UST owners/operators. First, insurance programs available at the time generally offered UST insurance up to $2 million, so financial responsibility levels within this limit would mean that existing insurance could be used to comply with the regulations.[18] EPA also reasoned that the lower aggregate levels would allow more UST owners/operators to use mechanisms other than insurance, such as guarantees, surety bonds, and letters of credit, to meet the financial responsibility requirements.[18] The reduced aggregates also would make it easier to capitalize risk retention groups (RRGs) and state funds to provide financial responsibility coverage.[18]

One of the most striking features of these changes in financial responsibility requirements is that for a very large and vulnerable class of tank owners — petroleum marketing firms with fewer than 12 tanks — they provide no real improvements over EPA's initial proposal. In fact, based on the reductions in coverage requirements, these changes provide the greatest benefits to those firms with the most USTs (and also with the greatest ability to provide evidence of financial responsibility).

Considering that firms owning only one retail outlet constitute 80% of all the firms in the retail motor fuel market, some UST owners/operators might be surprised that EPA made no changes in financial responsibility coverage levels from its initial proposal to benefit this category of tank owners.[23] In fact, the agency had no choice but to maintain these levels since the HSWA required a minimum of $1 million in coverage for *all* petroleum marketing firms.[22] Thus, Congress would have to pass legislation amending the HSWA in order to authorize EPA to lower UST financial responsibility levels any further for petroleum marketing firms.

7.3.2.2 Financial Responsibility Compliance Schedule

The final UST financial responsibility regulations set the earliest compliance dates for those UST owners/operators most likely to already have adequate financial resources, while giving more time to groups generally lacking financial responsibility capability. The "phase-in" categories used to establish the compliance schedule are based on UST ownership for petroleum marketing firms and on networth for nonmarketing firms. The four original compliance dates and categories included:

- January 24, 1989: Category I — All petroleum marketing firms owning 1000 or more USTs and all UST-owning non-marketing firms with a tangible networth of $20 million or more
- October 26, 1989: Category II — All petroleum marketing firms owning 100 to 999 USTs
- April 26, 1990: Category III — All petroleum marketing firms owning 13 to 99 USTs at more than one facility

- October 26, 1990: Category IV — All petroleum marketing firms owning 1 to 12 USTs or owning only one facility with fewer than 100 USTs; and all UST-owning nonmarketing firms with a tangible net worth below $20 million.[1]

To prevent firms with USTs at many locations from considering each entity separately to qualify the individual operators for later compliance dates, the regulations require that "regardless of which party complies, the date set for compliance at a particular facility is determined by the characteristics of the owner."[24]

The compliance schedule for tank Categories I and II is relatively short because EPA thought that UST owners with more than 100 tanks generally already had — or could obtain — insurance that would meet the financial responsibility requirements or had sufficient net worth to qualify for the guarantee or financial test of self-insurance. The agency recognized that compliance by tank owners in the other categories would be more difficult.

Thus, the categories and dates were designed to "achieve the maximum balance between the need to ensure financial capability for UST releases and the necessary time for owners and operators to obtain assurance mechanisms."[25] EPA recognized that the task of ensuring compliance would be challenging — the agency estimated that "95% of UST-owning firms .... for which financial assurance must be provided do not currently have pollution liability insurance and would not qualify to use the financial test for self-insurance...."[26] As the compliance dates for Categories III and IV approached, however, it became apparent that the balance that EPA sought might not have been achieved.

7.3.2.2.1 Modification of Compliance Dates for Categories III and IV

EPA officials were forced to reassess the feasibility of small and medium UST-owning firms obtaining financial responsibility by their respective 1990 compliance deadlines. From the time the final regulations were passed, the agency had acknowledged that "insurance and state financial assurance programs are likely to be the most feasible mechanisms for most owners and operators to comply" with the financial responsibility regulations.[27] Later EPA was more specific in stating that the original deadlines were based on the idea that Category III firms would rely on insurance while Category IV firms would use state assurance funds.[28]

Unfortunately, the agency was too optimistic in its assumption that the insurance market would expand to provide adequate coverage and that EPA would expeditiously approve the state funds that were expected to allow small UST owning firms to meet the financial responsibility requirements. EPA estimated that less than 2% of USTs owned by Category III firms had UST insurance as of March, 1990.[28] Although estimates were not made of the percentage of Category IV firms with UST insurance, the figure probably approached zero. In addition, only nine state funds had been approved by the agency and most of these still required that tank owners obtain some level of insurance coverage, upgrade their tanks, or both.[28] As of January, 1992, EPA had approved 27 state funds, most of which still require UST owners/operators to maintain some additional form of financial responsibility.[3]

As a result of the agency's reassessment, and amid strong pressure from associations of tank owners, EPA announced March 15, 1990 that it would extend the compliance deadline for Category III and Category IV tanks by 1 year, leaving new compliance dates of April 26, 1991 and October 26, 1991, respectively.[30] It was hoped that the extensions would provide adequate time for these Categories to comply with the financial responsibility requirements. Apparently the agency was convinced in December, 1991 that the new deadline for Category IV owners/operators was too soon because it extended the financial responsibility deadline for this group to December 31, 1993.

7.3.2.2.2 Suspension of Enforcement

Since these changes in the program, EPA's extended deadline (April 26, 1991) for Category III firms has also passed. And as mentioned above, the previous deadline for Category IV firms (October 26, 1991) was recently postponed to December 31, 1993. Unfortunately, many of these firms still face severe compliance problems as the UST insurance industry continues to sputter despite EPA's expectation that it would expand in the face of the "captured market" of firms forced to meet EPA's financial responsibility deadlines. In addition, in most cases state funds remain only partial solutions at best to the needs of the smallest UST-owning firms.

To many observers, the most effective approach — from the standpoint of the regulated community — is to employ the provision in HSWA authorizing EPA to suspend enforcement of the financial responsibility requirements for a particular class or category USTs if the agency determines that methods of "financial responsibility .... are not generally available for such tanks and:[22]

- Steps are being taken to form a risk retention group for the class or category of tanks, *or*
- The effected state is taking steps to establish a fund as proscribed by the law."

Enforcement of the law can be suspended for 180-day periods; the suspension can be renewed after the situation is reviewed by EPA.

Before the original 1990 compliance deadlines for Category III and IV firms were reached, several members of the House of Representatives* (primarily

---

* A bill to extend compliance deadlines and provide loan guarantees to small UST owning business also was introduced in the senate. See S. 2175, 101st Cong., 2nd Sess.

from Texas, Kansas, Montana, Oregon, and Ohio), introduced a bill that would have suspended enforcement of the financial responsibility requirements "until EPA can certify that reasonably affordable insurance is available."[31] The bill was not enacted, but its introduction did indicate that the financial responsibility compliance problem was gaining political attention.

In its discussion of the final UST financial responsibility regulations, EPA indicated that "even with a phase-in compliance schedule, ... the suspension of enforcement provisions remain available to owners and operators who can meet the requirements of the suspension provision."[26] Although a possibility, UST owners/operators should not depend on EPA suspending enforcement of the financial responsibility regulations, because this is intended as a last resort. In fact, when the financial responsibility regulations were published, EPA stated that with the flexibility provided by the phase-in schedule, suspension of enforcement should *not* be used for situations where there is "inadequate time to .... meet insurers' preconditions."[25] And although there was some support in Congress to suspend enforcement of the requirements, this was before EPA extended the compliance deadlines, in effect responding to Congress' concern.

Enforcement of the UST financial responsibility regulations has not yet been an issue for most UST owners/operators since the April 26, 1991 compliance deadline for Category III firms (13 to 99 USTs) passed only recently while the compliance deadline for Category IV firms has been postponed. EPA issued final guidance to its regional officials for calculating penalties for violations of both UST technical and financial responsibility regulations on November 11, 1990.[32,33] The guidance requires regional enforcement officials to consider adjustments that could increase penalties over the previous method used to determine penalty amounts.

Factors that can be used to increase penalties over the amounts arrived at using the previous method include noncooperation with EPA, negligence, and violations of the law in spite of full knowledge of the requirements.[32-34]

Although they are not required to follow the EPA enforcement guidance, states that implement an UST regulatory program *in lieu* of EPA are required to provide for "adequate enforcement of compliance" with the regulations.*

In January 1990, the Petroleum Marketers Association of America expressed its concern to EPA over possible enforcement of the financial responsibility regulations for Category II firms (100 to 999 USTs) which were required to comply with financial responsibility requirements by October 26, 1989. The concerns raised by PMAA centered on the competitive disadvantages that these firms may suffer compared to the smaller firms with later compliance deadlines. PMAA suggested that EPA avoid these potential

---

* EPA regulations for approving state UST programs require that, in order to operate the program *in lieu* of EPA, states must have the authority to assess administrative penalties (or sue to recover civil penalties in court) of at least $5,000 per tank per day for each violation of the UST regulatory requiremetns. 40 CFR 281.41(a)(3).

inequities by extending the compliance deadlines for all UST owners/operators with less than 1000 USTs.[35]

As mentioned in the beginning of this chapter, PMAA recently asked EPA to establish a single compliance deadline of December 31, 1992 for all firms in Categories II to IV.[4] Although PMAA has warned since before the UST financial responsibility regulations were published that penalties for noncompliance could threaten thousands of UST-owning firms, EPA has strictly limited its enforcement action[5] against any Category I or II firms. Thus, despite concerns among the regulated community, enforcement of the financial responsibility regulations (including determining whether firms are in compliance with the regulations) has not been a top priority within EPA or the states, with most efforts being aimed at assisting UST owners/operators in achieving compliance.[36] However, the current de-emphasis on enforcement is primarily because of the high compliance rates by the larger UST-owning firms in Categories I and II, so smaller firms should not assume that noncompliance with the financial responsibility regulations will be met with a *laissez-faire* attitude by either EPA or the states.

Regardless of the current attitude of EPA or the states on the enforcement of the financial responsibility regulations, the safest course for UST owners/operators should be to assume that the regulations will be enforced unless the enforcement provisions are suspended. Thus, although suspension of enforcement does not seem to be on the horizon, the uncertainty surrounding insurance availability and the adequacy of state funds as a financial assurance mechanism could mean that this provision is the only hope for many UST owners/operators.

In the meantime, however, owners/operators currently without an acceptable financial assurance mechanism have difficult choices to make. Undoubtedly, many in the regulated community will find it financially infeasible — or undesirable — to continue operating USTs. And unfortunately, for those that decide to comply with the financial responsibility regulations, each of the available options has significant drawbacks.

7.4 COMPLIANCE PROBLEMS IN THE REGULATED COMMUNITY

Since EPA first proposed regulations requiring that UST owners/operators maintain evidence of financial responsibility, it was widely known that there would be many problems facing the regulated community. These problems have been discussed and elaborated in varying degrees of detail, and their exact nature and scope has changed since the initial proposal. What remains, however, is the fact that over 3 years after EPA first proposed its financial responsibility program, hundreds of thousands of firms currently using USTs remain unable to comply fully with the regulations.

Put most simply, the problem is that the vast majority of UST-owning firms either cannot afford a suitable financial responsibility mechanism or a mechanism is not available. As mentioned previously, many small and medium-sized firms must depend on only two of the nine mechanisms currently allowed by EPA to meet the financial responsibility requirements — (1) insurance (or risk retention group) and (2) a state fund (or other state assurance). Both of these financial responsibility mechanisms have not developed as EPA intended. In addition to the issues of availability and affordability, concerns increasingly are being raised over the longterm feasibility of insurance and state funds as financial responsibility mechanisms for existing UST owners/operators.

7.4.1 UST Insurance

In many ways, the difficulty that UST owners/operators have obtaining insurance to cover the costs associated with leaks lies at the center of the debate over UST financial responsibility. There is little doubt that in 1988 EPA was overly optimistic in its projections that "12 to 18 months is a reasonable time in which to expect the insurance industry to respond to the increased demand for coverage" brought by the UST financial responsibility program.[37] In addition, the agency estimated that already there were several companies offering UST insurance, some to single-station "open dealers" and even to nonmarketers, and that several other firms were about to begin offering coverage.[37] As discussed in Chapter 9, however, most UST owners/operators already know that the supply of UST insurance has not kept up with its demand.

7.4.1.1 Limited Insurance Availability

Almost 2 years later (and 3 months after the original April 1990 compliance deadline for Category-III UST firms), a survey by the U.S. General Accounting Office (GAO — an auditing agency for Congress) provided strong evidence contradicting the scenario established by EPA in 1988. In fact, GAO found that out of 11 companies identified in the survey as offering UST insurance, 3 provided "more than 90 percent of the approximately 3,400 policies issued nationally" to firms in Categories III and IV.[10]

Perhaps more important than the number of UST insurance companies, the number of UST insurance policies being written is almost insignificant compared to the number of firms in the regulated community. To understand just how little coverage 3,400 policies represents for these smaller UST categories, a draft regulatory analysis conducted for U.S. EPA in 1986 estimated that 78,619 out of 79,657 firms (99%) in the retail motor fuel market owned less than ten retail outlets.[13] There are also tens of thousands of farms and small businesses outside of the retail motor fuel market that own USTs, many of which have a tangible net worth below $20 million, and thus are included in Category IV. In addition to the limited coverage provided by the three "largest"

UST insurance providers, the GAO study found that five of the identified companies had issued ten or fewer policies.[10]

A more recent study by the PMAA provided a slightly brighter picture of the UST insurance market, although the market was still not issuing policies commensurate with the number of firms in the regulated community. While only 15% of the smaller firms (i.e., less than 100 USTs) that were surveyed had UST insurance, the actual percentage of all firms in the regulated community with UST insurance undoubtedly is much lower.[38,39] This is because the smaller firms surveyed in the PMAA study owned an average of 34 USTs, substantially more than the 1 to 12 USTs owned by Category IV firms which form the bulk of the regulated community.[38,39] And Category IV firms generally have much less access to UST insurance than firms with substantially more USTs.

In addition, the study found that there still were few UST insurance companies in 1990, with three insurers providing 76% of the policies issued to surveyed firms.[38,39] Clearly, the number of insurance companies offering UST coverage and the total amount of policies being written remain severely limited.

7.4.1.1.1 The Rising Costs of UST Insurance

Besides the difficulty of qualifying for coverage, many UST owners/operators who *can* obtain coverage have found that premiums are rising sharply. In many cases, the cost of UST insurance is diverting funds from tank upgrading programs. This has led some members of the regulated community to call for more time to upgrade tanks before financial-responsibility regulations are enforced.

Comparison of PMAA surveys of premiums in 1989 and 1990 indicates a steady increase in the costs of UST insurance. Respondents to one survey reported spending an average of $27,969 on UST insurance in 1989.[40] This figure increased to $46,000 in the study of UST insurance premiums in 1990.[38,39] Perhaps more important for smaller UST-owning firms, the costs of UST insurance is much higher on a per-tank basis for firms with few USTs. In a 1990 PMAA survey, firms with less than 100 USTs paid an average premium of $1774 per UST, while those with more than 100 USTs only paid an average of $586.[38,39] Of the 73% of firms without any UST insurance in the 1990 survey, 46% said that the reason they lacked coverage was that it was unaffordable.[38,39] And unfortunately, because most UST leaks cost well below $1,000,000 — or even $500,000 — to remedy, lowering the amount of financial responsibility required might not significantly reduce premiums unless the amount is cut to (or below) the costs associated with an "average" leak. Clearly, the UST insurance industry still has a long way to go in providing the scope of coverage envisioned by EPA.

7.4.1.2 The Insurance Industry Perspective

The extremely limited availability and increasing cost of insurance coverage for Category III and IV firms is due mainly to the conditions of their USTs and the stringent criteria set by insurance companies to issue policies. These criteria are used because firms offering Environmental Liability Insurance (EIL) have experienced increasing "exposure" to liability and shifting regulatory conditions that make insuring USTs a risky business venture.

UST insurance policies usually are issued based on a careful individual assessment of risk and many insurance companies reject firms with contaminated sites, inadequate leak detection, unprotected bare steel tanks, or USTs over 15 years of age.[41-43] Unfortunately, many USTs Category III and IV firms exhibit one or more of these conditions.

7.4.1.2.1 *A Changing UST Insurance Climate and Lender Liability*

According to a 1988 report by the GAO, in the past UST insurers were attracted to the market largely because of two important factors that may not exist today (at least not to the extent they once did):[11]

- A large number of small and medium risks (cleanups typically averaging less than $30,000 to $100,000)
- Losses usually limited only to corrective action (few, if any, third party claims)

The 1990 PMAA study mentioned above revealed that 61% of surveyed firms reported conducting at least one tank/site cleanup.[38,39] Thus, given that the entire universe of USTs containing petroleum is undergoing a major regulatory "shake-up", and many — perhaps hundreds of thousands — leaking USTs will be discovered, it is not surprising that potential insurers would want to choose only the very best risks.

Some insurance officials have asserted that the UST insurance business is less attractive because of increased involvement by state and federal officials in groundwater protection activities — including oversight activities and corrective action regulations designed to ensure appropriate levels of cleanup. This involvement could lead to delays and increased corrective action costs. In fact, the PMAA study concluded that almost three fourths of the firms that had conducted a corrective action reported bureaucratic delays in the cleanup process. For each marketer that reported conducting corrective actions, an average of $168,000* was expended for site assessment and cleanup activities.[38,39]

In addition to problems with corrective action, increased discovery of groundwater contamination has heightened public concern and increased the role that third party liability plays in the total costs associated with an UST release.[11] Finally, as discussed in the following section, the use of state financial assurance funds could also deter the development of the UST insurance market.

Obviously, insurance companies have their own "schedule" for insuring USTs, and unlike EPA's, this schedule generally requires UST owners/operators to upgrade tanks and clean up sites *before* they obtain insurance. EPA's

---

* This figure includes amounts spent by firms with multiple LUSTs.

schedule, on the other hand, requires the regulated community to obtain financial responsibility mechanisms *first* and to upgrade tanks *afterwards* based on a 5 to 10 years schedule. It is this conflict in the requirements of EPA with those of the insurance industry that lies at the center of the UST insurance problem.

Given that many firms unable to obtain UST insurance must first perform certain tank upgrades, the need for loans becomes crucial to continue operating in compliance with the financial responsibility regulations. Unfortunately, many such firms seeking credit for tank upgrades find themselves caught in a "catch-22." Recent court cases have held banks liable for cleanup costs if the bank "had the capacity" to influence a borrower's decisions on how to handle hazardous waste.[44-49] Thus, bank loans made to firms later found to have caused environmental contamination exposed the bank to the liability for cleanup costs.

Although it is not clear that such "lender liability" would apply to loans made for tank upgrading purposes, many banks claim that the recent decisions have made them extremely reluctant to lend money for any businesses handling hazardous materials.[44-49] Of the UST-owning firms responding to the 1990 PMAA survey discussed above, 20% reported having had loan requests turned down by financial institutions based on liability concerns.[38,39] Thus, many UST owners/operators may be required to have clean sites and upgraded USTs to obtain insurance but will not be able to borrow the necessary funds if they do not have clean sites or sound USTs.

UST insurance — the only private mechanism that provides the necessary financial-responsibility levels without requiring UST owners/operators to have the money themselves — clearly will not be an option in the foreseeable future for many members of the regulated community. As a result, increasing attention has focused on state funds and other state programs comply with the financial responsibility regulations. Many different state programs have been developed, and EPA is accelerating its efforts to approve state funds so that UST owners/operators can comply with the financial responsibility deadlines. But because state financial responsibility programs often are little more than publicly subsidized insurance programs, it remains to be seen whether they can sustain compliance with EPA's financial responsibility requirements over the long term. (See Chapter 9 for a more detailed discussion of the use of insurance as a means to comply with federal UST financial responsibility regulations.)

7.4.2 State Financial Assurance Funds

Despite the emphasis placed on insurance as a means of demonstrating financial responsibility, EPA has recognized since 1988 that most UST owners/operators in Category IV (representing the majority of all regulated firms) would have to rely on state funds to comply with the financial responsibility requirements.[25] Thus, most firms in the regulated community probably will depend on state government — not the private insurance market — to comply with the financial responsibility program. But, as discussed in Chapter 10, the development of state funds that meet the needs of both the regulated community and the public has proven difficult, and EPA guidance on the matter has not eliminated the possibility that state financial assurance funds will encounter many problems.

Thus, as with private insurance, it appears that EPA may have overestimated the assistance that state funds would provide to the regulated community. As of April 1991, 46 states had established a fund to assist UST owners/operators in complying with federal financial responsibility regulations, and EPA has approved [27] of these funds for use as financial responsibility mechanisms.[3,29] Even though these funds currently are available to thousands of firms, it remains to be seen whether they will provide the regulated community with adequate coverage.

Perhaps the biggest threat to the use of state funds, however, is not that they will be shunned by states or disapproved by EPA, but that they rapidly may become both financially and politically infeasible and deter the development of a strong UST insurance market. Not surprisingly, the primary source of criticism of state financial assurance funds is the private UST insurance industry which views poorly-designed state financial assurance funds as a barrier to a successful UST insurance market. Thus, unless state funds can operate on sound financial principles, enjoy good political support, and co-exist with the private insurance market, it is not likely that they will provide a longterm financial responsibility option for UST owner/operators.

Where state funds cannot meet their financial responsibility obligations, eligible UST owners/operators could be exposed to liability for their leaking USTs or eventually could lose coverage from the fund altogether.[50] Therefore, UST owners/operators interested in ensuring the development of workable state financial assurance funds should understand those funds' limitations, along with their advantages.

7.4.2.1 The General Role of a Successful State Financial Assurance Fund

Many smaller firms with reasonably sound USTs, clean sites, and good tank management practices that are unable to obtain private insurance may be well served by certain types of state financial assurance funds. State funds, however, only provide UST owners/operators with a feasible compliance option if the fund is designed to minimize coverage for previous contamination and to provide adequate incentives for tank upgrading and good UST management practices. In addition to allowing firms to *comply* with the financial responsibility regulations, state funds also must be adequately capitalized and administered to *protect* the UST owner/operator from the liability associated with tank failure. Unfortunately for many UST owners/operators, a state fund meeting these criteria will resemble private insurance as much as — if not more than — a truly "public" assumption of the tank owners/operators' liability.

For firms with high risk USTs or contaminated sites that cannot afford UST upgrading, state funds will provide more limited benefits. Separate programs can be developed to assist UST owners/operators in cleaning up sites and upgrading USTs, but such programs probably would require huge levels of public financial and administrative support if they were applied to all firms unable to afford the transition on their own. In any event, state funds that provide financial responsibility and state programs that assist in UST upgrading and site cleanup should not be confused, as each has its own purpose and operating requirements.

In addition, a state fund should not necessarily be considered an *alternative* to private insurance. These funds may serve best as a temporary "stop-gap" mechanism to be used between EPA's compliance deadlines and the availability of private insurance. Continued participation in the state fund could prevent the development of the UST insurance market because of the inherent competitive advantage a subsidized state fund enjoys over a private insurance company that depends primarily on revenues from premiums in order to operate. In fact, this phenomenon of competition between state funds and private UST insurance already is taking place.

According to the PMAA studies, those firms surveyed with UST insurance had declined from 38% in 1989 to 17% in 1990, while coverage under state funds had increased from 16% to 57%* Of the surveyed UST-owning firms without insurance coverage in 1990, two-thirds stated that the primary reason for not obtaining insurance was their ability to meet the financial responsibility requirements using some other means (primarily state funds).[38,39]

A state fund also could deter insurers from providing coverage since the fund could adversely affect the factors (such as UST age, condition, and management) that insurance companies use to assess the risks involved in providing coverage to a particular area or specific category of UST-owning firms. Finally, state funds may not be designed to be compatible with any available UST insurance. Differences in eligibility requirements and UST management procedures are examples of potential areas of conflict.

Clearly, a permanently depressed UST insurance market would not be in anyone's best interest because the industry generally is better equipped than the state to reduce costs and promote a safer UST universe. UST insurance also is needed by UST owners/operators using state funds that only provide partial coverage for UST releases. These funds may only cover corrective action (not third-party compensation) or cover only cleanup costs above a certain amount, which can be as high as $200,000.

Despite these caveats, even a carefully designed state fund is subject to potential conflicts in its dual role of administering public funds and protecting

---

* As discussed in greater detail in Chapter 10, most of these state funds have not been approved by EPA. In addition, most state funds do not provide UST owners/ operators with all of the coverage they need to comply with the federal UST financial responsibility regulations.

human health and the environment. These conflicts are discussed in the following section.

7.4.2.2 Potential Conflicts in Administering State Funds

State financial assurance funds that assume overly broad liability for USTs can only be sustained under two circumstances: (1) very high levels of public funding are used for corrective action, third party compensation, and/or UST upgrading; or (2) state officials drastically reduce costs in cleaning up UST releases and compensating damaged third parties. In fact, the likelihood that such high levels of funding will not be available has left state financial assurance funds open to the criticism that states may be forced to limit clean-up and compensation activities in order to maintain the state fund.

Such a conflict of interest could occur in two ways. First, the state may hold that the fund cannot be used for either corrective action or third-party compensation for a particular UST release because the owner/operator violated some requirement for fund coverage, such as following specific UST management procedures. This holding would leave an unattended petroleum spill and possibly injured third parties, but as discussed below, would not benefit the state financially to the same extent as it would a private insurance company.

An insurance company can deny coverage for a spill caused by an UST owner/operator that fails to comply with its underwriting requirements since the damages caused by the spill do not affect the insurance company. After all, the insurer is not charged by the public as the protector of its natural resources. In many cases, however, a state that denies coverage of the financial assurance fund to an owner/operator for poor UST management would then have to turn around and expend different state funds to clean up the site to protect public health and the environment, whereas an insurance company would not be obligated to take any additional action. Thus, whether state resources are expended on corrective action for a firm covered by a state fund, or for cleanup in cases where a firm has violated conditions for coverage (and does not have adequate resources to pay for corrective action), the public — not the polluter — pays for the problem.

The second potential state conflict stems from the states' duty to prevent misuse of the fund by damaged third parties. For example, the state might wish to deny third parties access to the fund by claiming that they were negligent or not damaged — a common occurrence in the private insurance field. Again, however, the state has a legitimate interest in and responsibility for protecting both the fund and injured third parties. In addition to this conflict, if the state adjudicates the third party claim through the private tort system, it ultimately allows the judicial system to directly allocate funds supposedly managed by state government.

7.4.2.3 Specific Problems With State Funds and the Impact on Private Insurance Availability

Many of the general problems with state financial assurance funds discussed in Section 7.4.2.2 often are revealed by the way a particular state fund is structured and administered. The problems center around three specific funding elements: (1) the level of funding available and the way states finance the fund; (2) eligibility requirements; and (3) claims management. In addition, some 13 states have attempted to reduce the financial burden placed on state financial responsibility funds by establishing low-interest loan programs[13] to upgrade tanks before they are covered by the fund.* As discussed in Chapter 10, however, states must design financial assurance funds and loan programs for tank upgrading to limit the public's liability for UST leaks and provide UST owners/operators with adequate financial incentives to engage in sound UST management practices. Otherwise, these public approaches to demonstrating financial responsibility for UST owners/operators could prove to require unrealistic levels of public financing.

In addition, the very elements of state financial assurance funds that threaten their own longterm feasibility also could reduce the supply of UST insurance — the only other feasible financial responsibility mechanism for most UST owners/operators. (See Section 7.4.2.2.) Under this scenario (and in sharp contrast to the one offered by EPA in 1988), many state funds could falter over the next few years and simultaneously frustrate the development of a responsive UST insurance industry. This would leave many UST owners/operators in the same situation that they currently face — violating EPA's financial responsibility regulations with very limited options for reaching compliance. (See Chapter 10 for a more detailed discussion of the use of state funds as a financial responsibility mechanism.)

7.5 SUMMARY

EPA's UST financial responsibility program clearly has not met all of the expectations of EPA and, as a result, most firms with USTs remain uncertain whether compliance with the program will be possible. This uncertainty still persists, even though the agency twice has attempted to make compliance easier for some smaller, less profitable firms.

First, EPA set lower annual aggregate limits for UST financial responsibility in its final regulations than it had proposed originally in 1987. In addition, the final requirements reduced the per-occurrence financial responsibility amounts from $1 million to $500,000 for UST-owning firms not engaged in petroleum marketing. In 1990, EPA again revised its financial responsibility

---

* These state loan programs also can assist UST owners/operators in qualifying for private UST insurance.

program to extend the compliance date by 1 year for most firms with under 100 USTs. One of those dates (for petroleum marketers with between 13 and 99 USTs at more than one facility) again has passed, but associations of UST owners/operators still report widespread inability to comply with the financial responsibility regulations.

Because of the requirements of the HSWA, EPA has not been able to reduce the $1 million per-occurrence amount of financial responsibility that firms engaged in petroleum marketing must demonstrate. And because corrective action and third-party compensation costs normally are well below $1 million, it is not clear that reduced coverage requirements would significantly reduce insurance premiums. In addition, reducing coverage requirements only to the bare-minimum required to respond to the "normal" UST leak would conflict with the "polluter pays" principle in cases of more severe contamination and third-party damages.

This principle enjoys strong support from the public, and from an economic perspective, provides firms with adequate incentive to *prevent* UST releases. In fact, one of the most prominent features of the UST portion of the HSWA is that the responsibility for UST releases rests with the tank owner/operator. Although there have been congressional attempts to assist UST owners/operators, so far Congress has not been persuaded to amend the HSWA and eliminate this reliance on the "polluter pays" principle.

Most firms attempting to comply with the UST financial responsibility regulations must rely on the two allowable options — insurance and state UST financial assurance funds — that are most likely not to require UST owners/operators to generate the full amounts of financial responsibility coverage solely from the operation of their businesses. Neither of these mechanisms have developed to the extent that EPA had hoped when it published its financial responsibility regulations. Understanding why they have not is important for any UST owner/operator who wants to know what ultimately may be required in order to comply with the financial responsibility program and the UST technical requirements.

Being a private economic entity, the insurance industry is free to respond to the UST financial responsibility program in the manner that best serves its interest. Because of three related conditions brought on by the HSWA and EPA's financial responsibility regulations, UST owners/operators should not be surprised that insurers have approached the situation cautiously, instead of rushing to the aid of firms seeking to insure their USTs:

- The hundreds of thousands of USTs that represent poor insurance risks that are expected to get insurance but that cannot afford prompt tank upgrades and site remediation

- The rapid increase in the number of leaking USTs discovered and the costs of cleanup brought on by EPA and state UST reporting and corrective action requirements

- The uncertain, often tenuous relationship between private insurers with state UST programs, especially with state financial assurance funds that insurance companies may not consider reliable "partners" in providing professional, efficient, and stable financial responsibility services.

Insurance companies have responded to these conditions by avoiding what they perceive as risky business. In doing so, essentially they have forced the regulated community to conform to the "polluter pays" principle which will require USTs that are not upgraded to go without insurance. They also have responded by raising premiums (at least in the short term) to reflect the increasing liability costs associated with environmental contamination in general, and the federal UST program in particular.

At least in the short term, state funds may allow many UST owners/operators to comply with the financial-responsibility regulations. But political pressure, brought on by the massive public financing required by certain types of funds, could force even these public mechanisms to reflect the "polluter pays" principle just as the free-market has maintained this principle in the UST insurance industry.

State funds that are under-capitalized, apply to all USTs, or provide firms with little incentive to upgrade, may not be feasible over time and may deter the development of a strong UST insurance market. Funds unable to meet their obligations due to inadequate capital reserves could shift liability for an UST release back on the responsible party or on an insurer providing coverage in conjunction with a state fund. In addition, the public might exert political pressure to restore the "polluter pays" principle to the implementation of the state fund. This could remove many firms from the coverage that the state provides.

Not all state funds are the same, however, and some funds are being developed that more closely resemble insurance in requiring clean sites and upgraded tanks as a requirement for coverage. If structured properly, such funds could be useful to tank owners that cannot obtain insurance simply because it is not being offered or is too expensive. But for firms with old, possibly leaking USTs or contaminated sites, state funds may provide only short-term relief. Whether they are covered under a state fund or forced to obtain insurance, the financial responsibility situation for these firms is perhaps best described with the saying: "there's no such thing as a free lunch."

Despite the gloomy outlook, UST owners/operators should continue searching for the coverage needed to comply with EPA's financial responsibility regulations. Insurance coverage is available for many firms, and the coverage offered, premiums, and underwriting requirements can differ significantly. In addition, no two state funds are alike and UST owners/operators should understand what their state is doing to assist them. In many cases, it may be appropriate for owners/operators to contact UST regulatory officials or state legislators with comments concerning possible improvements in the state UST financial responsibility program. Finally, EPA's UST financial responsibility regulations are very detailed in describing *how* an UST owner/operator must demonstrate financial responsibility. Therefore, it is important that *all* the regulations are understood before a particular financial responsibility mechanism is considered.

NOTES

1. 40 CFR 280.91.
2. Solid Waste Disposal Act Section 9006(d). These penalties may be assessed for violations of the UST *regulations*. In addition, UST owners/operators who violate an *order* by EPA to comply with the UST regulations can lead to civil penalties of up to $25,000 per day of noncompliance. Solid Waste Disposal Act Section 9006(a).
3. U.S. Environmental Protection Agency, Office of Communications and Public Affairs, Note to Correspondents, April 15, 1991.
4. Teri Richman, National Association of Convenience Stores, Phillip R. Chisolm, Petroleum Marketers Association of America, and Jeffrey L. Leiter, Society of Independent Gasoline Marketers of America, letter to William K. Reilly, Administrator, U.S. Environmental Protection Agency, January 30, 1991.
5. 40 CFR Part 280.
6. Hazardous and Solid Waste Amendments of 1984 (Public Law 98-616).
7. Superfund Amendments and Reauthorization Act of 1986 (Public Law 99-499), Section 205.
8. 52 FR 12664, 12793.
9. U.S. Environmental Protection Agency, Office of Underground Storage Tanks, Dollars and Sense, EPA/530/UST-88/005, Washington, D.C., December 1988.
10. U.S. Government Accounting Office, Underground Petroleum Tanks — Owners' Ability to Comply with EPA's Financial Responsibility Requirements, GAO/RCED-90-167FS, Washington, D.C., July 1990.
11. U.S. Government Accounting Office, Superfund — Insuring Underground Petroleum Tanks, GAO/RCED-88-39, Washington, D.C., January 1988.
12. U.S. Environmental Protection Agency, Underground Motor Fuel Storage Tanks: A National Survey, Washington, D.C., May 1986.
13. Sobotka and Company, Inc., Regulatory Impact Analysis for Proposed Technical Standards for Underground Storage Tanks, March 30, 1987.
14. Virginia State Water Control Board, Fiscal Year 1985 — Pollution Response Office Summary, (no date).
15. U.S Environmental Protection Agency, Summary of State Reports on Releases from Underground Storage Tanks, Washington, D.C., August 1986.
16. 52 FR 12786.
17. 53 FR 43322-43383.
18. 53 FR 43335-43338.
19. 53 FR 43324.
20. 40 CFR 280.93.
21. 40 CFR 280.92.
22. Solid Waste Disposal Act Section 9003(d)(5).
23. Meridian Research, Inc. and Versar, Inc., Financial Responsibility for Underground Storage Tank Releases: Financial Profile of the Retail Motor Fuel Marketing Industry Sector, Draft Interim Report Submitted to U.S. Environmental Protection Agency, Office of Underground Storage Tanks, May 22, 1986.
24. 40 CFR 280.90.
25. 53 FR 43331-43332.
26. 53 FR 10402.

27. 53 FR 43325.
28. 55 FR 18566-18567.
29. Faulkner, Barbara, Petroleum Marketers Association of America, memorandum to Association Executives, Motor Fuels Steering Committee, regarding Update on State UST Trust Funds, January 10, 1991.
30. EPA Extends Financial Assurance Deadlines For Certain Underground Storage Tank Owners, *Environmental Reporter,* March 23, 1990.
31. H.R. 3321, 101st Congress, 2nd Session.
32. U.S. Environmental Protection Agency, Office of Underground Storage Tanks, U.S. EPA Penalty Guidance for Violations of UST Regulations, November 1990.
33. U.S. Environmental Protection Agency, Office of Underground Storage Tanks, UST/LUST Enforcement Procedures Guidance Manual, May 1990.
34. First Guidance for Assessing Penalties for UST Regulation Violations Issued by EPA, *Environmental Reporter,* December 7, 1990.
35. Faulkner, Barbara J., Petroleum Marketers Association of America, letter to Ron Brand, U.S. EPA Office of Underground Storage Tanks, January 5, 1990.
36. Waldrop, Greg, U.S. Environmental Protection Agency, Office of Underground Storage Tanks, personal conversation with Paul Thompson, February 25, 1991.
37. 53 Federal Register 43345-43347.
38. Memorandum from Ethel S. Hornbeck, Petroleum Marketers Association of America, to Association Executives, Motor Fuel Steering Committee, and Operations and Engineering Steering Committee, regarding PMAA's 1991 UST Survey Results, February 8, 1991.
39. Hornbeck, Ethel S., 1991 Joint Survey of Gasoline Marketer Underground Storage Tank Activity, National Association of Convenience Stores, Petroleum Marketers Association of America, and Society of Independent Gasoline Marketers of America, January 1991.
40. Chisholm, Phillip R., Petroleum Marketers Association of America, memorandum to various PMAA committees, regarding Results of Pollution Liability Insurance Surveys, August 11, 1989.
41. Gulledge, William P., Developing a pollution liability insurance underwriting model: managing for the potential exposure from toxic releases, *The Environmental Professional*, Vol. 11, 277-290, 1989.
42. Gulledge, William P., Controlling environmental impairment from an insurance perspective: risk management for underground storage tanks, presentation at 81st Annual Meeting of the APCA, Dallas, Texas, June 19-24, 1988.
43. Metelski, John J., EIL insurance: handle with care, *Best's Review: Property/Casualty Insurance Division*, 87(11).
44. Efforts to Enact Lenders' Protection Will Continue, Bankers Vow, Superfund Report, November 7, 1990.
45. Fleet Factors: Amicus Brief of American College of Real Estate Lawyers, Superfund Report, November 7, 1990.
46. Fleet Factors: Amicus Brief by New England Legal Foundation, Superfund Report, November 7, 1990.
47. Liability Insurance — Lawmakers Seek Options for 1991 Superfund Law, Superfund Report, July 19, 1989.
48. U.S. Asks Supreme Court Not to Review Fleet Factors Ruling on CERCLA Liability, Environment Reporter, December 21, 1990.
49. U.S. House of Representatives, Committee on Small Business, Hearing on the Impact of Superfund Lender Liability on Small Businesses and Their Lenders, June 7, 1990.
50. Anderson, Myra A., State UST financial assurance funds: disasters waiting to happen, *Environmental Claims J.*, Spring 1989, 313-321.

CHAPTER 8

EPA's UST Financial-Responsibility Regulations

Paul Thompson

CONTENTS

0-87371-402-4/93/$0.00+$.50
© 1993 by Lewis Publishers

EPA's UST Financial-Responsibility Regulations

8.1 INTRODUCTION

Based on the authority established by the Hazardous and Solid Waste Act (HSWA) of 1984 and the Superfund Amendments and Reauthorization Act (SARA) of 1986, EPA published regulations on October 26, 1988 requiring owners or operators of USTs containing petroleum to demonstrate evidence of financial responsibility. The financial responsibility is required to cover the costs of corrective action and third-party compensation (including bodily injury or property damage) from sudden and nonsudden releases caused by the operation of USTs. The regulations for USTs containing petroleum are contained in the United States Code of Federal Regulations in Title 40, Part 280 (40 CFR 280).

Although EPA is charged with implementing the federal UST program (both the technical — and financial-responsibility regulations), the agency has made it clear that it expects states to "take ownership" of the program by establishing equally stringent state UST laws and regulations. Although only two states — Mississippi and New Mexico — have EPA-approved programs, UST owners/operators should be aware that many states eventually may have the responsibility for implementing the UST program.* But while state UST financial responsibility regulations may differ from those published by EPA, any variations should be minor because of the requirement that state UST programs be "no less stringent" than the federal program.** (See footnote on following page)

EPA's regulations cover a variety of issues associated with demonstrating financial responsibility and are broken into the following sections:

- Applicability (40 CFR 280.90)
- Compliance Dates (40 CFR 280.91)
- Definition of Terms (40 CFR 280.92)
- Amount and Scope of Required Financial Responsibility (40 CFR 280.93)

---

* Dave Hamnett, U.S. Environment Protection Agency, Office of Underground Storage Tanks, personal conversation with Paul Thompson, April 16, 1991. States may delegate authority for certain aspects of an UST program to local officials, such as building inspectors or fire marshals. Thus, state and local agencies authorized by EPA to implement an UST program are sometimes referred to in this chapter as "authorized agency".

- Allowable Mechanisms and Combinations of Mechanisms (40 CFR 280.94 to Section 280.102)
- Standby Trust Fund (40 CFR 280.103)
- Substitution of Financial Assurance Mechanisms by Owner or Operator (40 CFR 280.104)
- Cancellation or nonrenewal by a Provider of Financial Assurance (40 CFR 280.105)
- Reporting by Owner or Operator (40 CFR 280.106)
- Recordkeeping (40 CFR 280.107)
- Drawing on Financial Assurance Mechanisms (40 CFR 280.108)
- Release from the Requirements (40 CFR 280.109)
- Bankruptcy or Other Incapacity of Owner or Operator or Provider of Financial Assurance (40 CFR 280.110)
- Replenishment of Guarantees, Letters of Credit, or Surety Bonds (40 CFR 280.111)
- Suspension of Enforcement (40 CFR 280.112)

Obviously, compliance with EPA's regulations requires more than simply obtaining some form of a financial responsibility mechanism in the amounts mentioned in the previous chapter. This chapter provides a more detailed description of what UST owners/operators are expected to do in order to comply with the financial responsibility regulations. Not all of the requirements, such as some definitions and terms and the exact wording of certain documents used to demonstrate financial responsibility, are included in this discussion.

In addition to EPA's financial responsibility regulations, this chapter includes an Appendix that describes two other components of EPA's overall UST program. First, the use of the federal Leaking Underground Storage Tank (LUST) Trust Fund to clean up spills and leaks from USTs is discussed. In addition, the Appendix also describes the reporting and corrective action procedures that EPA requires UST owners/operators to follow when an UST leak is discovered. These procedures are important since they will have to be

---

** EPA also has published regulations that states must follow to have an UST program approved (40 CFR 281). State programs may receive *interim* or *final* approval from EPA. No states have applied for interim approval to implement an UST program. As mentioned in Chapter 7, only New Mexico and Mississippi had final approval from EPA to implement an overall UST program as of mid-1991. Four other states have completed applications for submission to U.S. EPA to seek final UST program approval, while approximately 15 others are working on applications. These applications could be submitted to U.S. EPA in the next 18 months. Jerry Parker, U.S. EPA Office of Underground Storage Tanks, personal conversation with Paul Thompson, April 19, 1991. Once a state UST program is approved, the state will have primary enforcement responsibility, although EPA retains the right to take enforcement action in approved states as necessary [40 CFR 281.11(c)(3)]. The regulations require that states authorize the use of civil penalties of up to $5,000 *per violation* for submitting false information under the UST notification requirements. Penalties of up to $5,000 *per tank, per day of violation* must be authorized for failure to comply with other UST requirements or standards [40 CFR 281.41(a)(3)].

followed when a particular financial responsibility mechanism, such as insurance or a state fund, is used to address a LUST.

The exact location of particular requirements in the Code of Federal Regulations (CFR) is given in parenthesis. In addition, any explanation of a particular requirement is followed by a volume- and page-reference to the Federal Register (FR) here the regulations were originally published and explained by EPA. Finally, Federal Register citations also are used for any amendments that have been enacted since the regulations were published on October 26, 1988, since most of these amendments have not yet been written into the Code of Federal Regulations.

8.2 APPLICABILITY (40 CFR 280.90)

All owners and operators of USTs containing petroleum that are subject to EPA's technical standards for USTs and that are in operation on or after the financial responsibility compliance dates also must comply with the financial responsibility regulations [40 CFR 280.90(a) and Section 280.90(b)]. Some clarification of "all owners and operators" is necessary. (USTs that are exempt from EPA's technical standards and financial responsibility requirements are listed under 40 CFR 280.90(d); compliance dates are discussed in Section 8.3 and under 40 CFR 280.91.)

State and federal governments that own or operate USTs are not required to comply with the regulations. EPA presumes that these entities already have adequate funds for taking corrective action and compensating damaged third-parties (53 FR 43327 to 43328). In addition, unless required to do so by the state or federal government, any private party that operates an UST for state or federal officials is not required to comply [40 CFR 280.90(c)].

As mentioned above, the financial responsibility regulations do not apply to USTs that are exempt from EPA's technical standards for UST construction, installation, operation, corrective action, and closure [40 CFR 280.90(d)]. The main exemptions are contained in the definition of USTs used in the Solid Waste Disposal Act (SWDA).* The UST definition is:

> any one or combination of tanks (including underground pipes connected thereto) which is used to contain an accumulation of regulated substances,** and the volume of which (including the volume of the underground pipes connected thereto) is 10 per centum or more beneath the surface of the ground.

---

* See Solid Waste Disposal Act Section 9001 (1).

** The term *regulated substances* includes materials other than petroleum products, but EPA's financial-responsibility regulations under 40 CFR 280 apply only to those tanks containing petroleum. The HSWA defines petroleum as "including crude oil or any fraction thereof which is liquid at standard conditions of temperature and pressure (60 degrees Fahrenheit and 14.7 pounds per square inch absolute)" Solid Waste Disposal Act, Section 9001(2)(B).

Specifically excluded from this definition are the following:

- Farm or residential tanks of 1000 gallons or less capacity used for storing motor fuel for noncommercial purposes
- Tanks used for storing heating oil for consumptive use on the premises where stored
- Septic tanks
- Pipeline facilities regulated under the Natural Gas Pipeline Safety Act of 1968 or the Hazardous Liquid Pipeline Safety Act of 1979 (or similar state laws)
- Surface impoundments, pits, ponds, or lagoons
- Stormwater or wastewater collection systems
- Flow-through process tanks
- Liquid trap or associated gathering lines directly related to oil or gas production and gathering operations
- Storage tanks situated in an underground area if the tank is situated upon or above the surface of the floor.

In addition, the regulations also contain many exemptions, including:

- UST systems containing hazardous waste*
- UST systems containing electrical equipment and hydraulic lifts
- UST systems used for wastewater treatment
- UST systems containing *de minimis* quantities of regulated substances, including those with a capacity of less than 110 gal, those serving as emergency backup tanks, sumps, field-constructed tanks, UST systems containing radioactive wastes and other radioactive materials
- Backup diesel tanks at nuclear facilities
- Airport hydrant fueling systems (53 FR 43329).

If owners or operators are different persons, only one has to demonstrate financial responsibility, but *both* parties are liable for noncompliance with the regulations. (Under EPA's technical regulations, both parties are responsible for corrective action in the event of a release.) Thus, the issue of which party demonstrates financial responsibility is left to the owner/operator. Owners/operators must obtain financial responsibility even if the previous tank owner or operator is responsible for any contamination at the site. EPA leaves the current provider of financial responsibility to seek relief in the tort system for recovering funds from previous contamination (53 FR 43326).

Regardless of which party demonstrates financial responsibility, the characteristics of the owner is used in determining the compliance date by which financial responsibility must be available [40 CFR 280.90(e)]

---

* Financial Responsibility regulations for USTs containing hazardous waste are either covered under Subtitle C of the Resource Conservation and Recovery Act or will be published later by U.S. EPA. U.S. EPA published an "Advance Notice of Proposed Rulemaking (ANPR) for financial responsibility regulations for hazardous waste on February 9, 1988", 53 FR 3818. These regulations have not been finalized.

8.3 COMPLIANCE DATES (40 CFR 280.91)

As mentioned in the previous chapter, the compliance dates for two categories of UST owners/operators were extended after final financial responsibility regulations were published. The following list includes the current compliance dates along with the volume and page of the FR notice that amended the final regulations.

- January 24, 1989 (Category I) — All petroleum marketing firms owning 1000 or more USTs and all other UST owners that report a tangible net worth of $20 million or more to the U.S. Securities and Exchange Commission (SEC), Dunn and Bradstreet, or the Rural Electrification Administration, 40 CFR 280.91(a).
- October 26, 1989 (Category II) — All petroleum marketing firms owning 100 to 999 USTs, 40 CFR 280.91(b).
- April 26, 1991 (Category III) — All petroleum marketing firms owning 13 to 99 USTs at more than one facility, 40 CFR 280.91(c). (This compliance date was officially established on May 2, 1990 in 55 FR 18566.)
- December 31, 1993 (Category IV) — All other UST owners not covered above, including (1) petroleum marketing firms owning 1 to 12 tanks (or 99 or less at only one facility), (2) local governments, and (3) all nonmarketing firms with a tangible net worth of less than $20 million. (This compliance date was established 56 FR 66369. Until EPA announced an extension in December 1991, the compliance date had been October 26, 1991.)

8.4 DEFINITION OF TERMS (40 CFR 280.92)

Definitions used in the regulations are not addressed in this discussion. Some definitions and terms that are important to understanding other parts of the financial responsibility regulations, however, are discussed in the appropriate following sections.

8.5 AMOUNT AND SCOPE OF REQUIRED FINANCIAL RESPONSIBILITY (40 CFR 280.93)

The regulations establish minimum *per-occurrence* and *annual aggregate* financial responsibility amounts for different classes of UST owners/operators. If a combination of financial responsibility mechanisms are used to cover different costs (for example, one mechanism may cover only corrective action and another third-party compensation), then each must provide the total amount of required coverage [40 CFR 280.93(d)]. The required levels of financial responsibility cannot be used to cover any legal defense costs that may result from an UST release [40 CFR 280.93(g)]. This exclusion of legal defense costs is significant because, in the past, many UST insurance policies have included these costs in the insurance coverage. In addition, these financial responsibility levels are federal requirements and do not limit the UST owner/operator's liability for costs beyond these levels that may arise under tort law or additional state laws* [40 CFR 280.93(h)].

- Per-occurrence amounts
    $1 million for owners or operators of USTs that are located at petroleum marketing facilities, or that handle an average of more than 10,000 gal of petroleum per month based on annual throughput for the previous calendar year.
    $500,000 for all other owners or operators. This includes all nonmarketing firms that handle an average of less than 10,000 gal of petroleum per month [40 CFR 280.93(a)(1)].

- Annual aggregate amounts
    $1 million for owners or operators of 1 to 100 USTs.
    $2 million for owners or operators of 101 or more USTs [40 CFR 280.93(a)(1)].

These annual aggregates apply to both petroleum marketers and to nonmarketers. The definition of UST used to set aggregate financial responsibility levels is "a single containment unit", and not combinations of tanks at an individual facility [40 CFR 280.93(c)]. If additional USTs are installed that move the firm from the $1 million to $2 million annual aggregate financial responsibility requirement, then the new $2 million of financial responsibility must be demonstrated on the anniversary of the date that the original financial responsibility coverage became effective [40 CFR 280.93(f)].

8.6 ALLOWABLE MECHANISMS AND COMBINATIONS OF MECHANISMS (40 CFR SECTION 280.94 TO SECTION 280.102)

Most of the UST financial responsibility regulations describe the types of mechanisms that can be used to demonstrate financial responsibility, including

---

* For example, an UST owner/operator would be liable for costs beyond the required financial responsibility amounts in tort cases brought by third parties for sums exceeding these required amounts. In addition, most states have the authority to order responsible private entities (including UST owner/operators) to clean up groundwater contamination and other natural resource damages. States also have the authority to determine the appropriate type of corrective action and level of clean-up. Finally, many states can assess civil and criminal penalties against private parties for environmental contamination in cases involving certain types of negligence or willful misconduct. In these circumstances, the federal UST financial responsibility program in no way limits the liability of UST owner/operator to the required levels of financial responsibility. UST owner/operators are still liable for these costs, even if they exceed the amount of financial responsibility required by EPA.

the information that must be contained in any required documentation and the financial criteria that must be met to use a particular mechanism. Although this section summarizes the most important aspects of these mechanisms, UST owners/operators considering the more complex mechanisms, such as the guarantee or trust fund, should obtain copies of the regulations themselves to more fully understand the requirements. Allowable financial-responsibility mechanisms include:

- Financial test of self-insurance
- Guarantee
- Insurance and risk retention group coverage
- Surety bond
- Letter of credit
- Use of state-required mechanism
- State fund or other state assurance
- Trust fund

8.6.1 40 CFR 280.95 — Financial Test of Self-Insurance

This financial responsibility mechanism was used by the largest petroleum marketing firms, particularly the major refineries. Self-insurance will not be a feasible option for most firms because the mechanisms required high-levels of financial resources. Owners/operators wishing to use this mechanisms must meet either of two "financial tests".

The first financial test requires firms using the mechanism to have a tangible net worth of at least 10 times the amount of aggregate financial responsibility required under the regulations, in addition to any other financial responsibility required under federal hazardous waste regulations.* Firms also must have a minimum financial net worth of $10 million. In addition, firms must file financial statements annually with the U.S. Securities and Exchange Commission, the Energy Information Administration, the Rural Electrification Administration, or report its annual net worth to Dunn and Bradstreet [40 CFR 280.95(b)].

The other financial tests require UST owners/operators to demonstrate the amount of liability coverage required for facilities that store, treat, or dispose of hazardous waste. This test allows firms with a tangible net worth of $10 million and at least six times the amount of required financial responsibility coverage to self-insure [40 CFR 280.95(c)].

If a firm discovers that it no longer meets either of these financial tests, it must obtain another suitable financial responsibility mechanism within 150 days of the end of the year for which financial statements have been prepared

---

* Facilities involved in the treatment, storage, and disposal of hazardous waste are regulated under Subtitle C of the Solid Waste Disposal Act (SWDA). The HSWA amended the SWDA by adding Subtitle I (Sections 9001 to 9010) to address USTs. Thus, petroleum contained in USTs is not regulated as hazardous waste.

[40 CFR 280.95(e)]. If EPA or the state agency responsible for the UST financial responsibility program requests financial information and discovers that the firm no longer qualifies as a self-insurer, the firm then has 30 days to obtain the new financial responsibility mechanism [40 CFR 280.95(f)].

8.6.1.1 The Use of Self-Insurance by Local Governments

Many local governments that use USTs to fuel municipal and county vehicles could not qualify for self-insurance because the financial test was designed for private corporations, particularly the net-worth requirement that failed to measure a local government's ability to meet financial responsibility obligations. In addition, most other allowable financial responsibility mechanisms, including guarantees and insurance, are not as available to local governments as private entities. Thus, four separate tests for local governments to use to demonstrate financial responsibility were proposed by EPA on June 18, 1990 (55 FR 24692 to 24709). EPA's final UST financial responsibility regulations for local governments were published in the summer of 1991.

Under EPA's proposal, local governments can use a bond-rating test, worksheet test, guarantee, or maintain a funded balance to demonstrate financial responsibility. The tests are similar in intent to the corporate guarantee (described below) and the financial test of self-insurance that can be used by corporation to demonstrate financial responsibility (55 FR 24693).

Under the bond rating test, the proposal would make local governments eligible for self-insurance if they had $1 million or more of total outstanding issues of general obligation bonds and investment-grade bond ratings. (Investment-grade bond ratings include Moody's rating of Aaa, Aa, A, and Baa, and Standard and Poor's ratings of AAA, AA, A, and BBB [55 FR 24697 to 24698].)

Those who do not qualify under the bond-rating test may use the worksheet test using several financial criteria designed to measure a local government's financial stability. Under the test, local governments would use financial data derived from 1982 census data to calculate nine ratios, including debt service to total revenue, total funds to total expenses, and debt service to population. These ratios are scored (based on comparisons with the averages for all municipalities) and weighted according to the significance of each. The total of all nine scores are then compared to a cutoff point established by EPA as the minimum acceptable level of financial stability. Those above that level would qualify to self-insure (55 FR 24699 to 24702).

A governmental guarantee also can be used to demonstrate financial responsibility. Under the proposal, a government acting as a guarantor must first pass the bond rating or worksheet test. Governments entering into a guarantee must demonstrate a "substantial business relationship." Under EPA's definition of a "substantial business relationship," municipalities could obtain a guarantee from a state, a town from the surrounding county, and a special district from the sponsoring local government entity (55 FR 24702 to 24703).

The fourth financial responsibility mechanism available to local governments is the maintenance of a fund balance. Under this option, the local government does not have to meet any measure of financial stability but must establish a dedicated fund that meets the requisite aggregate financial responsibility requirements. This fund requires the local government to establish irrevocable trusts (cash or investment securities) pledged to use in responding to catastrophic events, including UST releases. If the fund is used for general emergency response, then the combined fund balance must be at least ten times the required aggregate financial responsibility level for the locality (55 FR 24703 to 24705).

8.6.2 40 CFR 280.96 — Guarantee

UST owners/operators may obtain guarantees to demonstrate financial responsibility. Under a guarantee, the firm providing the guarantee (called the guarantor) essentially contracts with the state or EPA that, if the UST owner/operator fails to undertake required activities, the guarantor will provide the necessary funds from its corporate assets (53 FR 43345). To do so, the guarantor must meet one of two criteria. First, the guarantor must either possess a controlling interest in the UST owner/operator or be controlled itself through stock ownership by a common "parent" firm that also controls the UST owner/operator.* Second, the guarantor can be a firm engaged in a "substantial business relationship" with the UST owner/operator that issues the guarantee as a result of the relationship** [40 CFR 280.96(a)].

The guarantor must demonstrate that it meets the financial test for self-insurance within 120 days of the close of each financial reporting year and provide this information to the UST owner/operator. If the guarantor fails to meet the financial test, it must notify the UST owner/operator within 120 days of the close of the financial reporting year. If the guarantor is notified by EPA or the state that it no longer meets the financial test of self-insurance, the guarantor must inform the UST owner/operator within 10 days of the notification. In either case, the guarantee cannot terminate less than 120 days after the UST owner/operator receives notice that the guarantor failed to pass the financial test of self-insurance [40 CFR 280.96(b)].

---

* Controlling interest means direct ownership of at least 50% of the voting stock of another entity.

** The financial responsibility regulations define "substantial business relationship" as "the extent of a business relationship necessary under applicable state law to make a guarantee contract issued incident to that relationship valid and enforceable. A guarantee contract is issued 'incident to that relationship' if it arises from and depends on existing economic transactions between the guarantor and the owner or operator" [40 CFR 280.92(m)].

Although this mechanism could allow petroleum jobbers who supply open dealers or nonmarketers to provide guarantees for these smaller firms, it is doubtful that such an approach will be used extensively. First, the jobber would have to pass the financial test of self insurance, which is not possible for all but the largest petroleum suppliers. In addition, there is a trend among UST suppliers toward making customers financially responsible for leaks. This trend can be expected to continue, especially with the increased liability risks faced by firms engaged in potentially hazardous activities and the increased oversight of state and federal officials.*

8.6.3 40 CFR 280.97 — Insurance and Risk Retention Group

The bulk of the financial responsibility regulations for insurance and risk retention groups describe the policy terms that must be included in a valid insurance policy. These terms are worded to ensure that the policy conforms to the rest of EPA's UST financial responsibility requirements. Some of these requirements may not have been standard policy terms for UST insurance in the past, so UST owners/operators are cautioned to examine potential insurance policies closely.

The policy must cover on-site cleanup in addition to any contamination that leaves the premises of the policy holder.** This on-site coverage does not, however, include the costs associated with response action that are part of routine maintenance of the tank site, site restoration, or site enhancement. In addition to on-site coverage, acceptable insurance policies also must cover non-sudden accidental releases in addition to sudden releases. Although some insurance policies may not cover non-sudden releases, the HSWA requires that financial responsibility be available for this type of release. In addition, EPA has noted that it is often difficult to determine whether a release was sudden or non-sudden (53 FR 53348).

In addition to these differences between policy terms acceptable under EPA's financial responsibility regulations and those used in traditional UST insurance, there are other potential differences that the UST owner/operator should look for. These potential differences are in the coverage terms *occurrence, accidental*

---

* See Thompson, Paul S., Conn, W. David, and Geyer, L. Leon, *Financial Responsibility Provisions for Underground Storage Tanks in Virginia*, Virginia Water Resources Research Center, February 1988, p. 20.

** Insurance companies may be reluctant to offer coverage for on-site damages because this could provide a disincentive to the UST owner/operator to maintain his site properly, thus encouraging increased UST releases and insurance claims. Coverage for on-site damage (called first-party coverage) has traditionally been offered as a separate type of policy. Still EPA has retained the requirement that insurance cover on-site damage, noting that such early response can prevent more expensive corrective action and third-party compensation (53 FR 43348).

*release, and bodily injury.** (53 FR 43348 to 43351). UST owners/operators should ensure that any UST insurance policy reflects the terms as given in the regulations. These are

- Occurrence — An accident, including continuous or repeated exposure to conditions, which results in a release from a UST
- Accidental release — Any sudden or non-sudden release of petroleum from a UST that results in a need for corrective action and/or compensation for bodily injury or property damage neither expected nor intended by the tank owner/operator
- Bodily injury — The meaning given to this term by applicable state law; however, this term shall not include those liabilities which, consistent with standard insurance industry practices, are excluded from coverage in liability insurance policies for bodily injury (40 CFR 280.92).

8.6.3.1 Three Financial Responsibility Requirements That May Conflict with Standard Insurance Industry Practices

The regulations describing acceptable insurance coverage contain three requirements that also could conflict with standard insurance industry practice. First, the policy must require the insurance company to notify the insured UST owner/operator before the policy can be canceled. The regulations originally required that all UST owners/operators receive notice at least 60 days before the policy was canceled, but the regulations were amended on November 9, 1989 to require only a 10 day notice if the policy were being canceled due to nonpayment of premiums or misrepresentation of information on an insurance application. Policy cancellation for any other reason, however, still requires that the policy remain in effect for at least 60 days after the cancellation notice is received by the UST owner/operator (54 FR 47077 to 47082).

The second insurance condition requires the policy to provide "first dollar coverage." This means that the insurance company is required to pay for corrective action and third-party compensation within the amount of the deductible in case the UST owner/operator does not — or will not — contribute the deductible amount. This provision was designed to ensure that disputes between the insurance provider and UST owner/operator over who is responsible for paying amounts within the deductible would not delay site cleanup and third party compensation (53 FR 43349).

The third important insurance provision required by EPA is an "extended reporting period" of 6 months. Under this provision, UST releases that occurred while a "claims-made" policy was in effect must be covered for 6 months after the policy is canceled, thus giving the UST owner/operator the additional time to determine whether or not an UST leak occurred during the policy period. It should be stressed that this "extended reporting period"

---

* These terms apply to any mechanism used to demonstrate financial responsibility. The terms are defined in the regulations at 40 CFR 280.92.

requirement does not mean that leaks occurring after the policy was canceled must be covered, only that leaks occurring while the policy was in effect may be reported up to 6 months after the policy is canceled* (53 FR 43350 to 43351).

Since most insurance policies are renewed with retroactive dates that extend coverage back to the end of the previous policy period, EPA amended this requirement on November 9, 1989 to clarify that the extended reporting period was required only in situations where a gap in coverage would occur.** This clarification should have the effect of allowing UST owners/operators to purchase the extended reporting coverage at the end of the policy period (i.e., once they determine that it will be needed), rather than at the beginning, as was the typical insurance practice. The regulations now require UST owners/operators who will suffer a gap in their financial responsibility coverage to purchase extended reporting coverage before the end of their current policy period (54 FR 47077 to 47082).

Finally, as with all other allowable financial responsibility mechanisms, acceptable insurance policies must provide that the full amounts of coverage are available for corrective action and third-party compensation, *exclusive* of any legal defense costs that may be incurred. This could conflict with the standard practice of insurance companies that want to prevent high legal defense costs on top of the amount of funds it must pay in claims. UST owners/operators also should be aware that the regulations allow insurance companies to write policies that limit their liability for defense costs (53 FR 43351 to 43352)

8.6.4 40 CFR 280.98 — Surety Bond

UST owners/operators may obtain a surety bond to demonstrate financial responsibility as long as the company issuing the bond is listed by the U.S. Department of Treasury as an acceptable surety on federal bonds. Under the terms of a surety bond, the surety company is liable for the cleanup costs in instances where the UST owner/operator fails to perform as guaranteed by the

---

* As discussed in greater detail in Chapter 9, *claims-made* insurance policies typically cover only those releases that are reported during the policy period and that begin after the policy's retroactive date (usually the same date as the effective date of the policy). *Occurrence-based* policies cover a release that occurred within the policy period but these releases can be reported at any time. Most UST insurance is issued on a claims-made basis (53 FR 43349).

** Two situations that EPA identified that would require the purchase of extended reporting coverage are: (1) when an UST owner/operator renews an existing policy, or purchases a new policy with a retroactive date after the retroactive date of the previous policy; and (2) when a policy is terminated or not renewed and the UST owner/operator chooses to use a financial-responsibility mechanism other than insurance. Under this second situation, the extended reporting period would only be required if the termination or nonrenewal of the policy would leave a gap in coverage (See 54 FR 47077 to 47082).

bond. If the surety sustains a loss, it still has the right to take the owner/operator to court for restitution. Thus, unlike the insurance industry, surety companies issue bonds on the presumption that the principal (UST owner/operator) has adequate financial resources to cover any losses.*

Those UST owners/operators with a tangible net worth below $10 million that are interested in using a surety bond generally do not represent a very good risk, since many would have severe financial difficulty in sustaining losses up to $1 million. (In fact, the whole point of insurance is to have someone else (an insurance company) assume the risk that *can* absorb the loss.) Only if the UST-owning firm survived as a business could the surety company ever hope to recoup any of its losses in cases where a firm had refused — or been unable — to cover the costs of an UST leak itself. Thus, it is unlikely that surety bonds will be a feasible method for demonstrating financial responsibility (53 FR 53353).

8.6.5 40 CFR 280.99 — Letter of Credit

UST owners/operators may use a letter of credit obtained from an authorized lending institution to demonstrate financial responsibility. A letter of credit is an instrument through which a financial institution undertakes to meet a monetary obligation of its customer (in this case, the UST owner/operator). A fee is charged by the institution for this service. Thus, the letter of credit would allow the UST owner/operator to draw the required levels of financial responsibility from the institution in the case of an UST release that could not be covered by the UST owner/operator.

Like surety bonds, however, letters of credit probably will not be a viable option for most UST owning firms. Letters of credit would require high levels of collateral from UST owners/operators that may "approach or exceed the face value of the letter of credit", thus making this option too expensive for many firms. The costs of letters of credit also are much higher than insurance. Finally letters of credit are not made available to many local governments by some lending institutions and some city codes prevent governmental bodies from securing letters of credit (53 FR 43353–43354).

8.6.6 40 CFR 280.100 — Use of State-Required Mechanism

States without UST programs approved by EPA, or those that require UST owners/operators to demonstrate financial responsibility may establish state-required financial responsibility mechanisms to meet the federal financial responsibility requirements. Although most states still lack full EPA approval of UST programs, most states have only used this mechanism as a supplement

---

* See, Dennis E. Wine, Surety Association of America, letter to Docket Clerk, Office of Underground Storage Tanks, U.S. Environmental Protection Agency, April 8, 1988.

to other financial responsibility mechanisms, most commonly state funds. As discussed in Chapter 10, unless a state fund offers "first dollar" coverage for a LUST, the UST owner/operator is required to demonstrate financial responsibility (called "wrap-around coverage") for any deductible amounts that must be paid before the state fund is used. The state-required mechanism allows states to determine how financial responsibility for this deductible amount can be demonstrated, without UST owners/operators having to resort to the other mechanisms, such as insurance, required by the federal financial responsibility regulations.

For example, states that require UST owners/operators to pay deductibles before a state fund can be used at a site often offer a "financial-test" to allow owners/operators to demonstrate financial responsibility for the deductible amount.* Such a test would be designed to demonstrate that UST owners/ operators will have funds only to pay the deductible part of the state fund (usually between $5,000 and $50,000), not the entire $500,000 to $2 million financial responsibility levels. Thus, it would be illogical for a financial test established as a state-required mechanism to set net-worth limits equivalent to the federal financial test. (The financial test of self-insurance used to meet the entire financial responsibility levels is described above.)

The regulations for a state-required mechanism give EPA a good deal of flexibility in determining what is acceptable if the mechanism is designed only to demonstrate financial responsibility for deductible amounts. In order for state-required mechanisms to be approved, EPA must determine "that the state mechanism is at least equivalent to the financial mechanisms" in the federal regulations. EPA must evaluate the adequacy of the state-required mechanism based on three factors [40 CFR 280.100(b)]:

- Certainty of the availability of funds for taking corrective action and/or for compensating third parties
- The amount of funds that will be made available
- The types of costs covered

Thus, if a state were proposing a mechanism to be used as the *sole* means of demonstrating financial responsibility, EPA could evaluate the mechanism by comparing it with the dollar amounts and allowable mechanisms in the federal financial responsibility regulations. But since the federal regulations do not provide any objective criteria to judge the adequacy of mechanisms providing only *partial* coverage, EPA has considerable discretion in assessing state proposals.

For instance, the federal financial responsibility regulations clearly indicate that demonstrating the *full* amount of financial responsibility using the financial test requires a net worth of at least $10 million and between six and ten

---

* David Hamnett, U.S. Environmental Protection Agency, personal conversation with Paul Thompson, February 6, 1991.

times the required financial responsibility levels. The regulations do not, however, describe the financial characteristics that are required of firms demonstrating financial responsibility for a $10,000 deductible. In this instance EPA could approve state-required mechanisms that require UST owners/operators only to have a net worth that is *three* times the deductible level (e.g., a net worth of $30,000 for a $10,000 deductible).*

Petitions to EPA to approve a state-required mechanism may be submitted by the state, an UST owner/operator, or any other interested party. Pending a final determination by EPA, UST owners/operators using the state-required mechanism will be deemed to be in compliance with the financial responsibility regulations for the amounts and types of costs covered by the mechanism [40 CFR 280.100(e)].

8.6.7 40 CFR 280.101 — State Fund or Other State Assurance

The federal financial responsibility regulations allow states to demonstrate financial responsibility for UST owners/operators through a fund or some other mechanism. Under the regulations, EPA must evaluate the state funds based on the same criteria used to evaluate the *state-required mechanism* discussed in the previous section [40 CFR 280.101(b)]. A state fund may cover all or only a particular class of USTs in the state [40 CFR 280.101(c)].

After a state submits a description of the state fund to EPA for official approval, the UST owners/operators to be covered by the fund are deemed to be in compliance with the financial responsibility requirements for the amounts and types of covered costs, unless EPA decides not to approve the proposed state fund [40 CFR 101(c)]. (As discussed in Chapter 10, the official guidance used by EPA to review state funds allows for significant variation in the types of funds that the agency will approve for use as a financial responsibility mechanism. In addition, many state funds only cover a portion of the financial responsibility required by the federal regulations. Under these funds, UST owners/operators are required to meet their remaining financial responsibility obligations through another acceptable mechanism.)

Within 60 days after EPA notifies the state that the fund proposal has been approved, the state must provide each UST owner/operator that is covered by the fund a letter or certificate describing the financial responsibility that is being assumed by the state. This letter or certificate must include the name and address of the facility covered by the fund and the amount of funds for corrective action and third party compensation that is assured by the state. The UST owner/operator is required to maintain this letter or certificate on file as proof of the financial responsibility [40 CFR 280.101(d)].

---

* David Hamnett, U.S. Environmental Protection Agency, personal communication with Paul Thompson, February 6, 1991.

8.6.8 40 CFR 280.102 — Trust Fund

The UST financial responsibility regulations allow the use of trust funds to demonstrate financial responsibility, either alone or in combination with other mechanisms. Under this mechanism, a trustee would pay into the trust fund, which would serve as a depository for funds needed to cleanup UST releases. The trustee would then be obligated to provide funds as directed by EPA or an authorized agency for corrective action and third-party compensation costs associated with an UST release.

As with most other mechanisms, the trust fund is not a viable method of demonstrating financial responsibility for most UST owners/operators. Trust funds normally are more costly than other financial responsibility mechanisms, and at any rate, require some party to supply the trust fund with the requisite levels of financial responsibility (53 FR 43354 to 43355). Therefore, *unaffiliated* UST-owning firms would probably be required to make this contribution themselves, which would be impossible in many — if not most — cases.

8.7 STANDBY TRUST FUND (40 CFR 280.103)

The UST financial responsibility regulations require that a standby trust fund be used in conjunction with the guarantee, surety bond, and letter of credit. Funds drawn from these mechanisms for corrective action and third-party compensation first must be deposited into the standby trust fund [40 CFR 280.103(a)]. The standby trust fund is necessary because, otherwise, any funds drawn directly from the guarantee, surety bond, and letter of credit that were made payable to the state or EPA would go into the state or federal treasury, respectively, where they would not be available for immediate use at the UST site (53 FR 43355). Thus, the use of the standby trust allows such funds to be used specifically for UST corrective action and third-party liability claims. The wording required for the standby trust fund is the same as required for the trust fund described in Section 8.6.8 [40 CFR 280.103(b)].

8.8 SUBSTITUTION OF FINANCIAL ASSURANCE MECHANISMS BY OWNER OR OPERATOR (40 CFR 280.104)

This section allows an UST owner/operator to substitute alternate financial responsibility mechanisms as long as compliance with the UST financial responsibility regulations is maintained (including an effective financial responsibility mechanism or combination of mechanisms) at all times. A financial responsibility mechanism may be canceled by the UST owner/operator after another acceptable mechanism (or combination) is obtained.

8.9 CANCELLATION OR NONRENEWAL BY A PROVIDER OF FINANCIAL ASSURANCE (40 CFR 280.105)

Providers of financial responsibility mechanisms to UST owners/operators may cancel the mechanisms only after a notice of termination is sent to the owner/operator by certified mail. The regulations require that guarantees, surety bonds, and letters of credit cannot be terminated until 120 days after the notice of termination has been received by the owner/operator. Insurance, risk retention group coverage, and coverage under state funds cannot be terminated until 60 days after the owner operator has received notice of termination [40 CFR 280.105(a)].

If a financial responsibility mechanism is canceled for reasons other the incapacity of the financial responsibility provider, the UST owner/operator has 60 days to obtain a financial responsibility mechanism once the notice of termination has been received. (See Section 8.10, for requirements applicable to situations where a financial responsibility mechanism was canceled because of the incapacity of the financial responsibility provider.) If a new mechanism is not obtained within 60 days, the owner/operator is required to inform the EPA or authorized agency and submit (1) the name and address of the financial responsibility provider, (2) the effective date of termination, and (3) the evidence of financial responsibility subject to termination [40 CFR 280.105(b)].

8.10 REPORTING BY OWNER OR OPERATOR (40 CFR 280.106)

This section requires UST owners/operators to report to EPA or an authorized agency under certain conditions that evidence of financial responsibility is being maintained in accordance with the regulations. First, the owner/operator must submit evidence of financial responsibility within 30 days after the owner/operator identifies an UST release. In addition, evidence of financial responsibility must be reported within 30 days after receiving notice of

- Commencement of a bankruptcy proceeding under Title 12 of the U.S. Code that names a provider of financial responsibility as a debtor
- Suspension of revocation of the authority of a provider of financial responsibility to issue a valid financial responsibility mechanism
- Failure of a guarantor to meet the requirements of the financial test
- Other incapacity of the provider of financial responsibility [40 CFR 280.106(a)].

In addition, UST owners/operators using the self-insurance financial responsibility mechanism must report to EPA or the state within 10 days of the inability to obtain a valid financial responsibility mechanism if the mechanisms is not obtained within: (1) 150 days of discovering that the firm no longer meets the financial test of self-insurance; or (2) 30 days after notification by EPA or an authorized agency that the firm no longer meets the financial test [40 CFR 280.106(a)(3)].

UST owners/operators also are required to notify state or local agencies designated in the regulations when a new UST is installed. As part of that notification process, UST owners/operators also must certify compliance with the UST financial responsibility requirements [40 CFR 280.106(b)].

Finally, EPA or the state may require UST owners/operators to submit evidence of financial responsibility or any other information relevant to compliance with the financial responsibility regulations at any time [40 CFR 280.106(c)].

8.11 RECORD KEEPING (40 CFR 280.107)

The UST financial responsibility regulations require UST owners/operators to maintain certain records to verify that financial responsibility is being maintained in compliance with the regulations. The records must be maintained at the UST site or the owner's/operator's place of business. These records are the documents associated with the individual financial responsibility mechanisms specified in 40 CFR 280.95 to 280.102 (Sections 8.6.1 to 8.6.8). In addition, UST owners/operators using the financial test of self-insurance or guarantee must maintain on file a copy of the chief financial officer's letter based on year-end financial statements for the most recent completed financial reporting year. UST owners/operators covered by a state fund must have a copy of any evidence of coverage supplied or required by the state that operates the fund [40 CFR 280.107(b)].

All UST owners/operators also must maintain an updated copy of a "certification of financial responsibility" worded according to the regulations. The certification must be updated whenever the financial responsibility mechanism changes [40 CFR 280.107(b)(6)].

8.12 DRAWING ON FINANCIAL ASSURANCE MECHANISMS (40 CFR 280.108)

This section describes the conditions under which EPA or the state may order a guarantor, surety, or institution issuing a letter of credit to place funds into the standby trust. EPA or the state may take such action when an UST owner/operator fails to obtain an alternate financial responsibility mechanism within 60 days after receiving notice that the guarantee, surety bond, or letter of credit has been canceled and the EPA or state determines or suspects that the covered UST has leaked [40 CFR 280.108(a)(1)].

The standby trust fund also may be drawn on when EPA or the state determines that a leak has occurred and, after appropriate notice and opportunity to comply, the owner/operator has not conducted corrective action as required under the UST technical regulations [40 CFR 280.108(b)].

The standby trust fund can be drawn on when EPA or the state has received certification from the UST owner/operator and injured third parties (including their attorneys) that a third-party liability claim should be paid. In addition, the fund can be used when EPA or the state determines that the UST owner/operator has not satisfied a valid final court order establishing a judgment against the UST owner/operator for third-party damages [40 CFR 280.108(b)].

If EPA or the state determines that the amount available through the standby trust will not cover necessary corrective action and third-party compensation costs, priority will be given to corrective action [40 CFR 280.108(c)].

8.13 RELEASE FROM THE REQUIREMENTS (40 CFR 280.109)

This section releases the UST owner/operator from the financial responsibility requirement after the UST has been properly closed (see Part I), or if corrective action is required, after corrective action has been completed and the UST has been properly closed. The UST closure regulations are provided in 40 CFR 280.70 through 40 CFR 280.74, and the UST release reporting and corrective action regulations are provided in 40 CFR 280.50 through 40 CFR 280.67. (EPA regulations for reporting releases from USTs and performing corrective action are discussed in the Appendix to this chapter.)

8.14 BANKRUPTCY OR OTHER INCAPACITY OF OWNER OR OPERATOR OR PROVIDER OF FINANCIAL ASSURANCE (40 CFR 280.110)

If an UST owner/operator is named as a debtor in a bankruptcy proceeding under Title 11 of the U.S. Code, the owner/operator must inform EPA or the state of the situation within 10 days of the commencement of the proceeding and supply the appropriate forms documenting financial responsibility [40 CFR 280.110(a)]. Guarantors providing financial responsibility that are named as debtors in such proceedings must inform UST owners/operators of the situation within 10 days of commencement of the proceedings, in accordance with the requirements of 40 CFR 280.96 [40 CFR 280.110(b)].

UST owners/operators that do not use the financial test of self-insurance are deemed to be without a financial responsibility mechanism in the event of bankruptcy or incapacity of their financial responsibility provider (e.g., an insurance company). After receiving notice of such an event, the UST owner/operator has 30 days to obtain an appropriate financial responsibility mechanism [40 CFR 280.110(c)]. This same 30 day period applies to UST owners/operators that receive notice that a state fund is no longer capable of paying its specified levels of coverage for corrective action and third-party compensation [40 CFR 280.110(d)].

8.15 REPLENISHMENT OF GUARANTEES, LETTERS OF CREDIT, OR SURETY BONDS (40 CFR 280.111)

This section requires that if a standby trust fund that is supported by a guarantee, letter of credit, or surety bond is drawn down below the amounts of required financial responsibility to address a leaking UST, the UST owner/ operator must either replenish the financial responsibility mechanism to the required level or obtain another financial responsibility mechanism, such as insurance, to make up the difference. Such action must be taken by the anniversary date of the guarantee, letter of credit, or surety bond.

8.16 SUSPENSION OF ENFORCEMENT (40 CFR 280.112)

EPA may suspend enforcement of the financial responsibility requirements for "a particular class or category of USTs or in a particular state" if EPA determines that methods of financial responsibility are not generally available for USTs in that class or category, and

- Steps are being taken to form a risk retention group for the UST class or category; *or*
- Such state is taking steps to establish a state assurance fund to be used as evidence of financial responsibility.

The suspension lasts for 180 days and can be extended at the end of this period if EPA determines that (1) substantial progress has been made in forming a risk retention group or (2) such a group is not possible and the state is unwilling or unable to establish a financial responsibility fund.*

As mentioned in Chapter 7, although this authority is currently available to EPA, the agency has not yet published regulations pertaining to suspension of enforcement. Consequently, this section of the regulations remains reserved for such regulations, should EPA find the need to suspend enforcement of the financial responsibility regulations.

8.17 CONCLUSION

Although EPA's UST financial responsibility regulations contain many requirements applicable to UST owners/operators, most firms in the regulated

---

* See, SWDA Section 9003(d)(5)(D).

community should be concerned primarily with only three: (1)compliance dates (40 CFR 280.91); (2) amount and scope of required financial responsibility (40 CFR 280.93); and (3) allowable mechanisms and combinations of mechanisms (40 CFRs 280.94 through 280.102). Many owners/operators may find other parts of the financial responsibility regulations to be inconvenient, but these three requirements most likely will present the greatest challenges to most UST-owning firms.

Two compliance dates for those owners/operators with the most USTs have already passed. Petroleum marketing firms with more than 1000 tanks and other firms with a tangible net worth of more than $20 million (Category I firms) were required to demonstrate financial responsibility by January 24, 1989. Category II firms (those with 100 to 999 USTs) had until October 26, 1989 to comply with the financial responsibility regulations. Virtually all Category I firms have complied with the regulations, with most using the financial test of self-insurance to demonstrate financial responsibility.

Although there are no national estimates, the compliance rate among Category II firms probably is high, although not so high as in Category I. UST insurance for Category II firms, although expensive, is more available than for smaller firms and many Category II firms also have the financial resources to use other financial responsibility mechanisms, such as the financial test or guarantee. There has not been any indications of widespread noncompliance among Category II firms or of fines levied by EPA or the states — who have yet to vigorously enforce the financial responsibility regulations — for violations of the financial responsibility regulations .

UST owners/operators falling into Categories III and IV, however, have not reached nearly the level of compliance enjoyed by the larger UST-owning firms. The compliance deadline for Category III firms (those with 13 to 99 USTs) passed on April 26, 1991, even though it was widely believed that many, if not most, of these firms had yet to comply fully with the financial responsibility regulations. UST insurance and state funds generally are the only financial responsibility mechanisms available to Category III firms, although insurance is by no means widely available or affordable. In the last 2 years, coverage of Category III firms by state financial assurance funds has increased while the use of UST insurance may actually have decreased (partially in response to the availability of less expensive state funds).

Category IV firms (UST owners/operators with 1 to 12 USTs) must comply with the financial responsibility regulations by December 31, 1993 — more than 5 years after the regulations were first published. Available information suggests that compliance levels for these firms is very low. The availability of UST insurance remains very limited. Thus, Category IV firms are highly dependent on the use of state financial assurance funds as the mechanism with which to demonstrate financial responsibility.

Many of these state funds, however, still require UST owners/operators to obtain some additional form of financial responsibility to meet the full levels of financial responsibility required by EPA. At any rate, only 27 of 46* state funds had been approved by EPA as of early 1992. Again, the lack of emphasis placed on enforcement by EPA and the states should not be misinterpreted as a sign that firms in Categories III and IV generally are complying with the financial responsibility regulations — or that the regulations will not be enforced once the compliance deadlines pass.

All petroleum-marketing firms with USTs must demonstrate $1 million dollars in per-occurrence financial responsibility for UST leaks. Firms not engaged in petroleum marketing must demonstrate $500,000 in per-occurrence financial responsibility. *All* firms — including nonmarketing firms — with 1 to 100 USTs must demonstrate $1 million of annual aggregate financial responsibility. Those firms with more than 100 USTs must have $2 million of financial responsibility available annually.

Although EPA has attempted to ease the financial burden on nonmarketing firms with USTs (as well as small petroleum marketing firms), it is not clear that, as measured on a per-tank basis, these efforts will result in substantially lower compliance costs for smaller firms than for larger ones. The reduced per-occurrence levels for nonmarketers may not lead to lower insurance premiums because of the requirement that $1 million in annual coverage be provided and because the USTs owned and operated by nonmarketing firms are a greater insurance risk and thus, more costly to insure.

In addition, insurance premiums may not vary significantly for firms demonstrating only $500,000 vs. $1 million since most of the insurance companies' costs are incurred in cleanups below $500,000. And the only difference between the required financial-responsibility levels for small vs. large petroleum marketing firms is that the smaller firms only need to demonstrate $1 million in annual aggregate financial responsibility instead of the $2 million required of larger firms. Given that insurers consider smaller UST-owning firms a greater insurance risk and most small firms would not experience UST leaks requiring more than $1 million in financial responsibility, it is not surprising that, in many cases, UST insurance will actually cost *more* for smaller firms than for larger ones.

As mentioned previously, most UST owners/operators must depend on only two of the eight financial responsibility mechanisms allowed in EPA's regulations — insurance and state funds. Determining whether insurance is a viable financial responsibility option or if a particular state fund will provide the needed coverage may well be the most important decision an UST owner/operator makes with regard to the financial responsibility regulations. Obviously,

---

*   This figure is based on a comparison of 27 approved state funds out of 46 identified as of January 10, 1991. See, U.S. Environmental Protection Agency, Office of Communications and Public Affairs, Notes to Correspondents, April 15, 1991 and Barbara Faulkner, Petroleum Marketers Association of America, memorandum to Association Executives, Motor Fuels Steering Committee, regarding Update on State UST Trust Funds, January 10, 1991.

private insurance varies greatly in its availability, costs, coverage provided, premiums, and eligibility standards. And EPA guidance allows great variation in the type of assistance that a state fund may offer to UST owners/operators. The following two chapters explore the current status and long-term viability of each of these financial responsibility options, which have become so critical to the ability of most UST-owning firms to comply with the EPA's financial responsibility regulations.

APPENDIX

FEDERAL REGULATIONS FOR THE LEAKING UNDERGROUND STORAGE TANK TRUST FUND, RELEASE REPORTING, AND CORRECTIVE ACTION

As part of the federal response program for leaking USTs, SARA established a LUST Trust Fund to provide funds for corrective action under certain conditions.* The fund is financed by a tax on motor fuel (expiration date — December 31, 1991) and is authorized to be capitalized up to $500 million.

Before the UST technical regulations became effective (December 22, 1988), the fund could be used for corrective action when EPA or a state acting under EPA's authority determined that the corrective action was necessary to protect human health and the environment. After December 22, 1988, however, the use of the fund became restricted to the following situations:

- When, within 90 days or such shorter period as may be necessary to protect human health and the environment, no owner or operator can be found who is subject to the UST corrective action requirements who is capable of carrying out the corrective action properly
- When prompt action by the EPA or state is necessary to protect human health or the environment
- If the financial resources of the UST owner/operator, including the required amount of financial responsibility, are not adequate to pay the full costs of corrective action
- If the UST owner/operator has failed or refused to comply with an order to perform corrective action

The fund cannot be used for corrective action in circumstances where the UST owner/operator is in violation of UST financial responsibility regulations, except under the following circumstances:

- There is no financially solvent UST owner/operator
- Immediate action is required to an imminent and substantial endangerment of human health and the environment
- The necessary action involves temporary or permanent relocation of residents, alternative water supplies, or an exposure assessment, undertaken to protect human health

---

* This section is based on the Superfund Amendments and Reauthorization Act, Sections 205, Section 521, and Section 522.

EPA has the authority to recover all the funds expended from the LUST Trust Fund from an UST owner/operator who is not in compliance with the financial responsibility regulations. Otherwise, the Trust Fund is used only to pay for costs over the amount of required financial responsibility.

EPA can authorize a state to administer the Trust Fund by providing grants from the Fund if the agency determines that the state has the necessary capacity to issue corrective action orders, oversee cleanup activities, and recover funds from responsible parties. States generally are required to contribute 10% of the amount expended from the Trust Fund for corrective action.

While the LUST Trust Fund is an important source of emergency funds for cleaning up LUSTs, it does not meet the needs of UST owners/operators seeking compliance with EPA's financial responsibility regulations. First, it generally cannot be used in situations where an UST owner/operator is still financially solvent and in violation of the regulations. Even if such an owner/operator refused to comply with an EPA or state corrective action order, the money expended from the Fund can be fully recovered from the responsible party. Thus, UST owners/operators in states with money from the LUST Trust Fund should not consider reliance on the Fund as a means of avoiding EPA's UST financial responsibility regulations.

UST RELEASE REPORTING AND CORRECTIVE ACTION REGULATIONS (40 CFR 280.50 THROUGH 40 CFR 280.67)

In the event that an UST leaks, EPA has passed reporting and corrective regulations that require UST owners/operators to follow certain procedures to ensure that sites are cleaned up consistently and effectively. Thus, these regulations describe how the money made available by the financial responsibility mechanism must be used.

Although consultants who specialize in petroleum spills normally will coordinate the cleanup process, UST owners/operators should be familiar with these regulations since it is the owner/operator that is responsible for ensuring compliance with the all UST regulations, including those pertaining to corrective action. In addition, part of the corrective action regulations require that certain reporting procedures be followed when a spill or leak is discovered, which normally will be the direct responsibility of the owner/operator.

Violations of the corrective action regulations can lead to the same penalties as violations of the financial responsibility regulations — with civil penalties up to $10,000 a day per violation. Refusing to obey an order by EPA to comply with the regulations can result in civil penalties of up to $25,000 for each day of noncompliance with the order.

Release Reporting, Investigation, and Confirmation (40 CFR 280.50 through 40 CFR 280.53)

These regulations require owners/operators to take certain actions to report suspected releases from USTs, investigate and confirm whether a release has actually occurred, and initiate an appropriate corrective action when various types of spills have taken place.

Reporting of Suspected Releases (40 CFR 280.50)

UST owners/operators must report to EPA or an authorized agency within 24 hours (or another reasonable time specified by the implementing agency) any of the conditions described below. In addition, these conditions also obligate the UST owner/operator to follow the procedures for investigating and confirming releases. (Investigating and confirming UST releases is discussed in the following section.)

First, the discovery by the owner/operator (or any other party) of released regulated substances at the UST site or the surrounding area (such as the presence of free product or vapors in soils, basements, sewer and utility lines, and nearby surface waters) must be reported [40 CFR 280.50(a)]. Second, owner/operators must report any unusual operating conditions (such as the erratic behavior of product dispensing equipment, the sudden loss of product from the UST system, or an unexplained presence of water in the tank), unless the equipment is found to be defective but not leaking and is immediately repaired or replaced [40 CFR 280.50(b)].

Third, UST owners/operators must report any monitoring results from a release detection method required by EPA that indicate a release may have occurred.* Reporting is not required if: (1) the monitoring device is found to be defective and is immediately repaired or replaced; and (2) additional monitoring does not confirm the initial results. In addition, if UST monitoring is performed according to EPA's inventory control procedures, potential releases identified by this method do not have to be reported if a second month of data does not confirm the initial result.** [40 CFR 280.50(c)].

Investigation Due to Off-Site Impacts (40 CFR 280.51)

This section requires UST owners/operators to follow certain procedures (described in the following section) to determine if the UST system is the source of off-site impacts. These impacts include the discovery of regulated

---

* UST monitoring is required under SDWA Section 9003(c)(2). EPA regulations for monitoring are included in 40 CFR 280.43.

** EPA regulations for release detection using inventory control are provided by 40 CFR 280.43(a).

substances that has been observed by — or brought to the attention of — EPA or the authorized agency.

Release Investigation and Confirmation Steps (40 CFR 280.52)

Unless corrective action already has been initiated, UST owners/operators who observe the three conditions of a potential UST release (described above) must follow certain procedures to verify that a release occurred. These investigation and confirmation procedures must be completed within 7 days or another reasonable time specified by EPA or the authorized agency. Either of two sets of procedures may be used.

The first option for investigating and confirming a possible UST release is called the *system test* and involves conducting a "tank tightness test" in accordance with EPA's UST technical regulations.* UST owners/operators must repair, replace, or upgrade the UST system, and begin corrective action if the test indicates that the system, tank, or piping is leaking. If the test does not indicate that a leak exists, then further investigation is not required *as long as environmental contamination is not the basis for suspecting an UST release.* If environmental contamination is the basis for suspecting that an UST release has occurred, then UST owners/operators must conduct the second allowable release investigation and confirmation method, called a *site check* [40 CFR 280.52(a)].

The site check requires UST owners/operators to measure for the presence of a release where contamination is most likely to be present at the UST site. Sample types, sample locations, and measurement methods must take into account: (1) the nature of the stored substances; (2) the type of initial alarm for cause of suspicion; (3) the type of backfill; (4) the depth of groundwater; and (5) other appropriate factors for identifying the presence and source of the release. If the site check indicates that an UST release has occurred, then corrective action must begin in accordance with the UST corrective action regulations; otherwise, further investigation is not required [40 CFR 280.52(b)].

Reporting and Cleanup of Spills and Overfills (40 CFR 280.53)

This section requires UST owners/operators to "contain and immediately clean up a spill or overfill and report to the implementing agency within 24 hours" (or another reasonable time specified by the agency) in the following circumstances [40 CFR 280.53(a)]

- A spill or overfill of over twenty-five gallons of petroleum (or another reasonable amount specified by the implementing agency) or one that causes a sheen on nearby surface waters; and

---

* These regulations are provided in 40 CFR 280.43(c) and 40 CFR 280.44(b).

- A spill or overfill of a hazardous substance that results in a release to the environment that equals or exceeds its reportable quantity under the Comprehensive Environmental Response, Compensation, and Liability Act (CERCLA)*

In addition, UST owners/operators must contain and immediately clean up petroleum spills and overfills of less than 25 gal (and hazardous substance releases of less than reportable quantities). If the cleanup cannot be accomplished within 24 hours (or another reasonable time period established by EPA or the State), then the UST owner/operator must immediately notify EPA or the State [40 CFR 280.53(b)].

Release Response and Corrective Action for UST Systems Containing Petroleum or Hazardous Substances (40 CFR 280.60 through 40 CFR 280.67)

UST owner/operators with a confirmed UST release must comply with a variety of corrective action regulations. These are summarized below.

Initial Response (40 CFR 280.61)

After an UST release is confirmed, UST owners/operators must provide an initial response within 24 hours (or another reasonable time specified by EPA or an authorized agency). First, the release must be reported to EPA or an authorized agency. Immediate action must be taken to prevent any further release of the regulated substance into the environment. In addition, any fire, explosion, and vapor hazards must be identified [40 CFR 280.61(a)].

Initial Abatement Measures and Site Check (40 CFR 280.62)

Unless directed to do otherwise by EPA or an authorized agency, UST owners/operators are required to [40 CFR 280.62(a)]:

- Remove as much of the regulated substance from the UST system as is necessary to prevent further release to the environment
- Visually inspect any above-ground releases or exposed below-ground releases and prevent further migration of the released substance into surrounding soils and groundwater

---

* Reportable quantities of hazardous waste releases are given under 40 CFR 302. CERCLA Sections 102 and 103 require that releases of reportable quantities of hazardous waste be reported immediately (rather than within 24 hours) to the National Response Center. SARA Title III requires that such quantities be reported immediately to appropriate state and local authorities.

- Continue to mitigate any fire and safety hazards posed by vapors or free product that have migrated from the UST excavation zone and entered into subsurface structures (such as sewers or basements)
- Remedy hazards posed by contaminated soils as a result of release confirmation, site investigation, abatement, or corrective action activities. If these remedies include treatment or disposal of soils, the owner and operator must comply with applicable state and local requirements
- Measure for the presence of a release where contamination is most likely to be present at the UST site (unless this has already been done as part of a site check)
- Investigate to determine the possible presence of free product, and begin free product removal as soon as practicable in accordance (free product removal is discussed below)

In addition to these activities, UST owners/operators must submit a report to EPA or an authorized agency within 20 days of a release that summarizes the initial abatement measures described above [40 CFR 280.62(b)].

Initial Site Characterization (40 CFR 280.63)

This section requires UST owners/operators to provide EPA or an authorized agency with information regarding the UST site and the nature of the release, including information gained while confirming the release or completing the initial abatement measures. At a minimum, this information must include:

- Data on the nature and estimated quantity of the release
- Data from available sources and/or site investigations concerning the following factors: surrounding populations, water quality, use and approximate locations of wells potentially affected by the release, subsurface soil conditions, locations of subsurface sewers, climatological conditions, and land use
- Results of the site check required as part of the initial abatement measures
- Results of the free product investigation required as part of the initial abatement measures

This information must be submitted to EPA or an authorized agency within 45 days (or another reasonable period of time) of a release confirmation according to the schedule and format required by EPA or the authorized agency [40 CFR 280.63(b)]

Free Product Removal (40 CFR 280.64)

If the free product investigation indicates the presence of free product, UST owners/operators must remove it to the "maximum extent practicable" as determined by EPA or an authorized agency. Free product removal must take place while all other aspects of the corrective action continue. The free product removal process must conform to the following requirements:

- Free product removal must be conducted in a manner that minimizes the spread of contamination into previously uncontaminated zones by using recovery and disposal techniques appropriate to the hydrogeologic conditions at the site, and that properly treats, discharges, or disposes of recovery byproducts in compliance with applicable local, state, and federal regulations
- The minimum objective of the design of the free product removal system must be the abatement of free product migration
- Any flammable products must be handled in a safe and competent manner to prevent fires or explosions

In addition, unless directed to do otherwise, UST owners/operators must submit to the EPA or authorized agency within 45 days after confirming a release a free product removal report that provides at least the following information [40 CFR 280.64(d)]:

- The names of the persons responsible for implementing the free product removal measures
- The estimated quantity, type, and thickness of free product observed or measured in wells, boreholes, and excavations
- The type of free product recovery system used
- Whether any discharge will take place on-site or off-site during the recovery operation and where this discharge will be located
- The type of treatment applied to, and the effluent quality expected from, any discharge
- The steps that have been or are being taken to obtain the necessary permits for any discharge
- The disposition of the recovered free product

Investigations for Soil and Groundwater Cleanup (40 CFR 280.65

This section specifies the conditions under which UST owners/operators must conduct investigations of the release, the release site, and the surrounding area to determine the full extent and location of contaminated soils and the concentrations of dissolved product in groundwater. Such investigations must be conducted under any one of the following four conditions [40 CFR 280.65(a)]:

- There is evidence that groundwater wells have been affected
- Free product is found to need recovery under 40 CFR 280.64 (discussed above)
- There is evidence that contaminated soils may be in contact with groundwater
- EPA or an authorized agency requests such investigations, based on the potential effects of contaminated soil or groundwater on nearby surface water and groundwater resources

UST owners/operators must submit the information collected from the investigations as soon as practicable or in accordance with a schedule established by EPA or an authorized agency [40 CFR 280.65(b)].

Corrective Action Plan (40 CFR 280.66)

EPA or an authorized agency may order an UST owner/operator to submit additional information or to develop a corrective action plan for responding to contaminated soils or groundwater. Such an order may be issued at any point after EPA or the authorized agency reviews the information submitted in compliance with the regulations for initial response, initial abatement measures, or initial site characterization. (These regulations are discussed above.) The corrective action plan must provide for adequate protection of human health and the environment as determined by EPA or the authorized agency, and be modified, as necessary, to meet this standard [40 CFR 280.66(a)].

To verify that the corrective action plan will adequately protect human health and the environment, EPA or an authorized agency should consider the following factors [40 CFR 280.66(b)]:

- The physical and chemical characteristics of the regulated substance, including its toxicity, persistence, and potential for migration
- The hydrogeologic characteristics of the facility and the surrounding areas
- The proximity, quality, and current and future uses of nearby surface water and groundwater
- The potential effects of residual contamination on nearby surface water and groundwater
- An exposure assessment
- Any other information developed as part of the corrective action process

Once approved by EPA or an authorized agency, the plan must be implemented by the UST owner/operator. They must monitor, evaluate, and report the results of implementing the plan in accordance with a schedule and in a format established by EPA or the authorized agency [40 CFR 280.66(c)]. Owners/operators may implement the plan before it is approved if they first notify EPA or an authorized agency, comply with any conditions imposed by EPA or an authorized agency, and incorporate any self-initiated cleanup measures in the corrective action plan that is eventually approved [40 CFR 280.66(d)].

Public Participation (40 CFR 280.67)

The public participation section requires EPA or an approved agency to notify that portion of the public most affected by the release of any UST leak that requires a corrective action plan under 40 CFR 280.66 [40 CFR 280.67(a)]. Any site release information and decisions concerning the corrective action plan must be made available to the public for inspection on request [40 CFR 280.67(b)]. EPA or the authorized agency has the option of holding a public meeting before approving a corrective action plan to consider public comments on the adequacy of the plan. In addition, public notice must be given if it is found that implementing the corrective action plan will not achieve the cleanup levels established in the plan and, consequently, EPA or the authorized agency is considering terminating the plan [40 CFR 280.67(d)].

CHAPTER 9

The Use of Insurance as a Financial Responsibility Mechanism

W. David Conn

CONTENTS

0-87371-402-4/93/$0.00+$.50
© 1993 by Lewis Publishers

The Use of Insurance as a Financial Responsibility Mechanism

9.1 INTRODUCTION

The private insurance industry has a long tradition of providing individuals and firms with a means of "hedging" against uncertain future financial losses. By paying a "premium" to purchase an insurance policy, the individual or firm transfers the risk (less a "deductible") to an insurance company which pools the risks of many different policyholders. The total premiums collected must be sufficient not only to cover the company's losses but also to pay both its business costs and the profits which make it worthwhile to stay in business. To ensure adequate revenue, the company must rely on its ability to predict the *aggregate* cost and timing of future losses. A special kind of insurance company is a *risk retention group* which is formed by (and provides coverage to) individuals and firms in a common profession or industry.

There are potentially several advantages to the use of private insurance as a means of assuring financial responsibility for the owners of underground storage tanks (USTs). One is that policyholders are not required to commit capital (unless they belong to a risk retention group, in which case they usually make a one-time capital contribution). A second advantage is that, because policyholders together pay the full cost of any losses, the use of private insurance conforms (at least, in aggregate) to the "polluter-pays principle" which satisfies many people's sense of fairness.

A third advantage is that private insurance typically provides an incentive both to minimize risks (in advance of a hazard) and to minimize losses (once a hazard has occurred). The existence of this incentive may not be immediately apparent, since the very purpose of insurance is to insulate the policyholders from their possible losses; nevertheless, it is in the interest of *those providing the insurance* to persuade or require policyholders to adopt risk minimization measures and to act rapidly (and efficiently) to contain losses if/when they occur. In practice, because of their financial stake, insurance companies are often very diligent "enforcers" of technical requirements (more so, in some cases, than the regulatory agencies that are actually charged with enforcement responsibilities). The required payment of a deductible (in most cases) also provides an incentive for risk minimization by policyholders.

A fourth advantage of private insurance, some would argue, is that it provides a "free-market" solution to the problem of establishing financial responsibility, requiring a minimum of government involvement (beyond the setting and enforcement of appropriate rules and safeguards).

If insurance were available at an affordable price to all UST owners/operators, arguably there would be no need for state funds or other mechanisms to assure financial responsibility. Unfortunately, despite EPA's earlier optimism that insurance would become not only more widely available but also more generally affordable, to date this has not happened. The reality is that companies cannot be forced to offer affordable policies which meet the federal financial responsibility requirements, or at all.

This chapter reviews the history of environmental impairment liability (EIL) insurance (of which UST insurance is a special case) and EPA's requirements for UST insurance. It explores the problems of the UST market, and the relationship between insurance and state funds, and discusses what all of this means for owners/operators. Examples of insurance documents are provided, and the chapter ends with a review of Washington's unique program to promote the development of a private UST insurance market in that state.

9.2 HISTORY OF ENVIRONMENTAL IMPAIRMENT LIABILITY INSURANCE

Prior to the passage of the Hazardous and Solid Waste Act (HSWA), very few owners sought special insurance for their USTs. Until the 1960s, third-party risks of bodily injury or property damage due to pollution from USTs or other sources were potentially covered by Comprehensive General Liability (CGL) policies. Typically, these policies were *occurrence-based*, meaning that they continued to pay for losses resulting from occurrences during the policy period even though these losses might not be discovered until after (sometimes long after) the policy had lapsed. As awareness of the extent of pollution risks grew, companies sought to reduce their potential liability by limiting coverage to sudden and accidental releases instead of more gradual ones; they also shifted to *claims-made* policies which pay only on claims made during the policy period. Eventually, pollution risks were excluded altogether from most CGL policies. One reason for this exclusion was that courts had begun to decide cases involving pollution on the basis of *strict liability*, holding defendants liable regardless of any showing of negligence or failure to demonstrate reasonable care (which had been required under traditional common-law doctrine).

A more recent development was the emergence of EIL policies, designed specifically to cover third-party losses resulting from gradual pollution. The need for this coverage was accentuated by the enactment of the Resource Conservation & Recovery Act (RCRA) in 1976 and the Comprehensive Environmental Response, Compensation, and Liability Act (CERCLA [Superfund]) in 1980, both of which established regulations affecting hazardous substances and imposed financial responsibilities on entities that handle such substances. Despite substantial growth in the market for gradual pollution policies in the early 1980s, by mid-decade the number of insurers offering these policies had significantly declined because of "adverse trends inside and outside the insurance industry."[1]

Among the problems for insurers was the fact that courts had begun to subject them, under CERCLA, to very large retroactive liabilities for pollution damage and cleanup costs that (in their view) they had never intended to cover. Not only were losses at the time very large, but future losses became increasingly difficult to predict. Under these circumstances, the reinsurance market for gradual pollution policies virtually disappeared. An insurance industry publication in 1985 stated:[2]

> The pollution hazard has turned out to be much more complex and expensive than anticipated. Gearing up to underwrite this catastrophic risk, it is now realized, requires major commitments of time, dollars and specialized expertise which typically is not available on-staff.

This adverse experience with environmental impairment liability coverage was fresh in the minds of the insurance companies when federal financial responsibility requirements for USTs were introduced on a discretionary basis in the Hazardous and Solid Waste Amendments (HSWA) in 1984 and subsequently mandated in the Superfund Amendments and Reauthorization Act (SARA) in 1986.

9.3 REQUIRED FEATURES OF UST POLICIES

In order to meet the requirements of EPA's financial responsibility regulations, UST insurance policies must satisfy the following provisions:

1. They must, at a minimum, provide the required limits of liability. These limits may be less than those specified in EPA's regulations *only* if insurance is used in combination with some other financial responsibility instrument *and* the instruments in aggregate meet or exceed the limits.
2. The liability limits must be *exclusive* of legal defense costs, i.e., the costs of investigating, contesting, defending, or appealing against a requirement or claim imposed on the insured. This provision differs from that found in most EIL policies; typically, defense costs are counted against the liability limits. However, the regulations do not prevent insurance companies from placing their own separate limits on defense costs. Some specify a dollar limit, while others agree to continue paying defense costs as long as the policy's limits on liability for corrective action and third-party damage have not yet been reached.
3. UST policies must provide coverage for *on-site* corrective action as well as third-party liability. Again, this is not typical of EIL policies. Some insureres have argued that providing on-site coverage is a disincentive to tank owners to adequately maintain their tanks, but (as previously discussed) it does not stop the insurers themselves from putting pressure on the owners (e.g., by refusing coverage to those who fail to meet technical standards, perform

appropriate tests, etc.). On the other hand, some insurers do not object to paying for on-site corrective action since it is seen as a way of minimizing off-site liability, which is typically more costly. In any event, insurers do not provide (nor do EPA regulations require them to provide) coverage of activities that are considered to be the day-to-day responsibility of tank owners/operators or "normal" business expenses. Activities that are not covered include routine testing and maintenance, repair, and replacement of tank systems.

4. Deductibles ranging from $5,000 to $500,000 per occurrence are typically offered. However, regardless of the deductible selected, the policy must provide for the insurance company — and not the tank owner/operator — to pay the *first dollar*. In other words, when there is a legitimate claim, the insurance company is liable for the claim in its entirety (up to the applicable limits), with a contractual right to recover from the policyholder any amount paid within the deductible. The government's purpose is to reduce the chance that a delay in responding to a release might result from a dispute over the payment of a deductible. Although this increases their risk (in that the deductible may prove difficult to recover), some insurers prefer having first dollar responsibility since it allows them to exercise complete control over the response to a release. A disadvantage is that it may make insurers more wary of issuing policies to tank owners/operators with limited financial resources, and thereby further limit the market (especially to smaller businesses).

5. Insurers are required to offer an *extended reporting period* to follow the termination (for whatever reason) of a claims-made policy. This means that an occurrence taking place during the policy period can be reported up to 6 months afterwards. Insurers normally charge a substantial extra premium (typically, 50% or more of the original premium) for this provision. Despite some initial confusion about this requirement, it has now been established that the extension need not be requested (nor paid for) until it is actually needed (that is, when coverage is to be ended).

6. So that tank owners/operators may have an opportunity to find an alternative method of establishing financial responsibility, insurers are required to provide *60 days notice of cancellation*, except when the cancellation results from nonpayment of the premium or misrepresentation by the insured (in which case, under a revised rule, only 10 days are required).[3]

7. Insurers are required to attach an *endorsement or certificate of insurance* worded in accordance with EPA specifications.

9.4 THE RELUCTANCE TO PROVIDE UST INSURANCE

As previously mentioned, following promulgation of the financial-responsibility regulations, insurance companies (with certain exceptions) were reluctant to enter the UST insurance market. An important consideration was prior adverse experience in general with pollution liability insurance, as already discussed. Insurers worried that the courts might reinterpret the terms of their UST policies in unanticipated ways, leading to unexpectedly high payouts in the future.

Insurers were also faced with the absence of a *loss history* on which to base predictions of future losses. Consequently, they had little or no basis on which to set premiums. What limited past experience existed suggested that most claims for corrective actions would probably fall below $500,000, and that relatively few claims for third-party damages should be expected. The latter was in part due to the fact that contamination by petroleum products is usually fairly easy to detect (by smell or taste) before it does serious damage. On the other hand, several factors could be expected to drive up the losses, including the facts that the new regulations would cause people actively to look for problems and that, at least in the short term, there might be a scarcity of skilled services to address leakage problems (causing prices to increase due to the interaction of supply and demand). In any event, there was considerable uncertainty about the frequency and magnitude of future claims, aggravated by uncertainty about how courts would interpret the new policies (e.g., whether limitations on liability would be upheld).

9.4.1 Availability of UST Insurance

These and other problems common to EIL, as well as concerns about the time and investment needed to develop expertise in the specialized UST insurance market, resulted in a lack of insurers coming into the market, as mentioned in Chapter 8. The 1990 General Accounting Office (GAO) report stated that tank insurance was being offered for sale in most states, even to firms in Categories III (petroleum marketers owning 13 to 99 USTs at more than one facility) and IV (petroleum marketers owning 1 to 12 USTs and others, as previously defined). However, these policies were being offered by only a very small number of insurers, and they were being purchased by a very small proportion of Category III and IV tank owners/operators.[4] According to the *1991 Joint Survey* conducted by the National Association of Convenience Stores (NACS), the Petroleum Marketers Association of America (PMAA), and the Society of Independent Gasoline Marketers of America (SIGMA), the number of marketers having private insurance actually decreased during the 1989-90 year, although the number having some kind of financial assurance simultaneously grew from 59 to 79%, largely as the result of coverage by state funds.[5]

UST policies to date have been more readily available to petroleum marketers than to nonmarketers. This is because insurers generally consider that marketers, whose business revolves around petroleum, are more likely to pay close attention to their tank systems and to know what it takes to run them properly; marketers are also thought to have the greatest financial incentive to manage their tanks for minimum product loss.

In the early part of 1990, the UST insurance market was dominated by two companies, Federated Mutual Insurance Company and the Petroleum Marketers Mutual Insurance Company (Petromark). These companies had issued 1800 and 1000 policies, respectively, to Category III and IV tank owners/operators.[6] Federated offered UST coverage *only* to owners/operators who purchased the company's General Liability Insurance, and subsequently announced that it would no longer provide coverage in 14 states having trust funds to pay for tank cleanup.

Petromark was a risk retention group that offered group membership and insurance to any owner/operator (including nonmarketers) whose tanks were in compliance with applicable federal and state technical requirements. A one-time cash contribution, comparable to the annual insurance premium, was required to secure membership. Petromark was licensed to sell insurance under Tennessee regulations, which specified that the company must maintain financial reserves sufficient to fully cover the future value of current claims. Unfortunately, because of the uncertainties involved in predicting this future value (as previously discussed), Petromark's actuaries had great difficulty in agreeing upon a single number. Ultimately, having identified a likely range of $0 to $60 million, they settled on a "mid-range" figure of $32 million, which was larger than the company's assets at that time. Under these circumstances, the Tennessee authorities had no choice under the law but to discontinue the company's license to sell insurance.

Petromark's reinsurers had been Underwriters at Lloyd's of London. Lloyd's had only recently re-entered the EIL market, having previously left this market as the result of losses under CERCLA. By the time that Petromark's operations were terminated, Lloyd's had gained enough confidence about UST-related claims to offer primary coverage to most of Petromark's former customers. At the time of writing (May 1991), Lloyd's has completed a year of underwriting UST policies and has agreed to continue doing so for at least another year.

According to the GAO report, as of March 1990 a third insurer (the Environmental Impairment Purchasing Group) had issued 320 policies to Category III and IV tank owners/operators as well as 45 policies to nonmarketers. Another (the American International Group) had issued 200 policies each to Category IV owners/operators and nonmarketers. A few other companies were offering full or partial coverage to these smaller firms, and some had provided a number of quotes, but they had thus far issued only a few policies.[7] Subsequently, the *Joint Survey* of NACS, PMAA, and SIGMA reported that, as of January 1991, three companies had come to dominate the UST insurance market: Lloyd's (accounting for 29% of those insured), Federated (22%), and Agricultural Excess (25%).[8]

9.4.2 Nature of UST Policies

Because of the large number of tanks, insurers typically do not conduct site visits before underwriting new policies. Instead, they require prospective policyholders to provide detailed site-specific information, specifically:

1. Release potential characteristics, such as
   - Tank age
   - Corrosion protection controls
   - Leak detection controls
   - Tank design
   - Soil corrosivity
2. Site vulnerability factors, such as
   - Depth to water table
   - Soil hydraulic conductivity
   - Surrounding population
   - Aquifer use
   - Proximity to surface waters[9]

Although insurers may require that tanks meet "all applicable" tecnnical standards, this may not mean a great deal (for existing tanks, at least) until the new federal requirements are phased in over the next few years. In the meantime, insurers often refuse coverage for unprotected steel tanks and/or any tanks over a specified age (15 to 25 years). They may require precision testing of the tanks, to be performed by a company either of the insurer's choosing or from a list of approved testers. In addition, they may require site audits, soil borings, etc., to establish the extent (if any) of preexisting contamination. This contamination may be the result of accumulated spills, for example, during filling operations. It is typically excluded from coverage.

According to insurance company representatives who responded to GAO's 1990 survey, for firms in Categories III and IV, average policy premiums per tank ranged from $800 to $1600.[10] Respondents to the *Joint Survey* (in which the smallest firms may be under-represented) reported late in 1990 that the average premium cost per tank had increased from $1369 to $1415 during the previous year. The trade associations' report pointed out that "insurance costs are not only escalating, they are also increasingly regressive. While the premium cost per tank to the largest marketers actually declined slightly to $586, smaller marketers report spending nearly $1800 per tank this year compared with around $1500 1 year ago."[11]

One of several factors affecting premiums is level of deductible. GAO's 1990 report described minimum deductibles as ranging from nothing to $25,000. Much larger deductibles ($50,000, $100,000, or $150,000) are now being offered on Lloyd's policies, issued through The Planning Corporation, as a way of allowing less stringent underwriting requirements. Such deductibles, however, are typically infeasible for small firms, which may therefore be precluded from purchasing these particular policies. The larger per tank premiums often paid by smaller firms may reflect the fact that they commonly require smaller deductibles than the larger firms.

As mentioned earlier, virtually all UST policies are issued on a "claims-made" basis. Occasionally, a situation arises that results in the voluntary re-opening of a file by an insurer after termination of the policy period (and any applicable extended reporting period). As the result (for example) of a contractor's error or omission, or the state's initial failure to unambiguously define its specifications for corrective action, costs may be incurred that were not reported at the appropriate time. Insurers typically maintain loss reserve accounts in anticipation of such "IBNR" claims ("incurred but not reported"), but they may contest them if it is suspected that pre-existing damage was discovered during cleanup of the new release.[12]

A specimen UST insurance policy and other related literature issued by insurers are included (for illustrative purposes only) at the end of this chapter in the Appendix.

9.5 RELATIONSHIP BETWEEN INSURANCE AND STATE PROGRAMS

It has been argued that most states, in developing their UST financial responsibility programs, concentrated on providing an alternative to private insurance coverage rather than focusing on the "real need" to eliminate barriers to greater insurance participation.[13] Differences in fund characteristics (e.g., how and at what level they are financed, how they determine eligibility, how they handle reimbursement, and what criteria they use to guide cleanup) complicate the insurers' task in attempting to work with them. So-called "wrap-around" policies have been developed for most states, but the biggest problem, according to one insurer, is that

> in many instances, the insurer is asked to give up the customary right of site intervention necessary to minimize environmental damage and subsequent restoration costs . . . Of particular concern is that most state fund programs are being managed by environmental regulatory agencies, which are chronically under-staffed and which have no experience in insurance claim handling.[14]

Some state funds cover the *first portion* of the costs, leaving tank owners/operators in need of insurance (or some other means) to demonstrate financial responsibility for the remainder of the costs, up to the federally specified limits. The problem of establishing management responsibility is the most severe for these kinds of funds. Other state funds cover the *excess* of costs over a specified amount; however, since the costs associated with most incidents have typically been less than $100,000, the limiting of exposure by a state fund may have little impact on the primary insurance (affecting premiums, for example, only slightly). A few state funds are *comprehensive* in their coverage (that is, they promise to pay all required costs); with these, insurers are unlikely to be competitive.

As mentioned in Chapter 7, there are concerns about the ability of many state funds to deliver all that they promise. Some, for example, appear to be severely undercapitalized. "Insurers fear they will be held liable for those costs intended for the state program but for which the state is unable to meet its responsibilities."[15] Unlike state funds, whose terms and conditions can be changed more or less at the whim of state government, insurance policies establish a contractual relationship between the insurer and the insured, at least for the period of the policy. "State funds", according to one insurer, "are intended to protect health and the environment, not necessarily the owners/operators; they may find themselves being asked after the incident to contribute to the costs involved."[16]

9.6 WHAT DOES THIS MEAN FOR OWNERS/OPERATORS?

As discussed in Chapter 7, it is the smaller firms, typically those in Categories III and IV, that are faced with the greatest difficulty in meeting the financial responsibility provisions of RCRA Subtitle I. Many of the larger firms in Categories I and II can meet these requirements through self-insurance, or are viewed as relatively "good risks" by the insurance industry (and are therefore readily insurable).

Data on the number of Category III and IV firms presently holding insurance coverage on their USTs may be a little misleading since many may have waited even to explore this option until their mandatory financial responsibility compliance dates (which for those in Category III was in April 1991, while Category IV firms are not required to comply until later in 1991). Nevertheless, it is apparent that many small firms will find it extremely difficult or impossible to obtain insurance, for several reasons:

1. Many own old tanks, some more than 25 years old
2. Many cannot afford to upgrade their tanks to meet the technical specifications demanded by the insurers (or even to perform the tests needed to establish eligibility)
3. Even when eligibility can be established, many cannot afford to pay the premiums charged by the insurers
4. The premiums are relatively high, on a per-tank basis, primarily because small firms typically require small deductibles (whereas large firms are more likely to be able to tolerate large deductibles)
5. Small firms tend to be relatively unsophisticated in dealing with technical and/or financial matters, and therefore they require considerable support from their insurers.

Some firms lack the option of going out of business, since they cannot afford even to meet the rules for tank closure. It is not surprising, therefore, that many small firms are counting on state funds to help them meet the financial responsibility requirements.

It is evident that insurance does not and cannot provide a panacea for UST owners/operators having difficulty in meeting both the financial responsibility and the technical requirements of Subtitle I. The most "difficult" cases are those owners whose USTs are not in compliance with the technical requirements, are currently leaking (or likely to do so at any time), but who lack the will and/or the financial means to undertake the necessary upgrade (and corrective action, if needed). Contrary to what some may have supposed, insurance is not intended to (nor will it) pay to correct existing problems. In order to qualify for insurance, if problems are likely or have already been identified, an owner must perform the upgrade and corrective action *first*. As previously mentioned, this requirement, needed to obtain financial responsibility, has the effect of accelerating the time of bringing USTs into compliance with the technical standards (which, in the law, need not occur until after the deadlines for financial responsibility compliance). It is generally recognized that some marginal operations are unlikely to be able to meet the requirements under any circumstances, and there is no reason to expect that insurance can solve the problems of these owners.

A related problem, mentioned in Chapter 7, is that some owners have found themselves facing a "chicken-and-egg" dilemma: they need to obtain a loan in order to upgrade their tanks prior to qualifying for insurance, but prospective lenders are unwilling to grant them a loan if they do not already have insurance. The problem of *lender liability* has arisen in part because of some lenders' adverse experience under the provisions of CERCLA; however, the liability provisions of RCRA Subtitle I differ in significant respects from those of CERLA, giving reason to suppose that courts might be less likely to impose unanticipated liabilities for UST-related damage and cleanup costs (at least for USTs containing petroleum products) than they have for other kinds of pollution costs. The term "liability" is defined identically in both statutes, suggesting that courts will decide that liability under Subtitle I is also strict, joint, and several. However, Subtitle I provides for third-party defenses, discretion in the recovery of costs incurred by the LUST Trust Fund, and limitations on potentially responsible parties that differ from provisions in CERCLA. Nevertheless, there remains considerable uncertainty about potential losses, and the perceived risks have made many lenders wary of establishing financial relationships with the owners/operators of USTs.[17]

There is some perception that even when insurance is obtained, and at an adequate level, an owner/operator's potential financial problems may not be over. There have been numerous occasions when insurance companies have contested EIL claims, especially in situations involving CERCLA cleanups. According to a 1991 GAO document, 13 insurance companies reported a total of 49,947 pollution claims open at the time of their response to a survey, not all of which would necessarily be closed with payment. The same insurers also reported that they were engaged in 1,962 lawsuits with insureds over pollution coverage issues (some of which may relate to the same contaminated sites).[18] An insurer has commented that most of the contested cases developed when EIL policies were evolving, for example, from "occurrence-based" to "claims-made" (as previously discussed). The language used in many of the earlier EIL policies made interpretation difficult, according to this insurer, and the sums involved (in CERCLA disputes) were very large. The GAO data apparently did not relate to UST insurance, and more recent policies have been written in a manner that allegedly is less susceptible to differences in interpretation.[19]

9.7 EXAMPLE OF A STATE REINSURANCE PROGRAM: WASHINGTON STATE

As mentioned in Chapter 7 (and discussed in greater detail in Chapter 10), most states have established funds to assist UST owners/operators in complying with the financial responsibility requirements of Subtitle I. The approach of Washington State is different from that of most others in that it is designed to stimulate an *insurance* market by having a government agency provide *reinsurance*. This section provides a brief description of the program in Washington, based largely on a recent article by a staff counsel who played a major role in its development.[20]

9.7.1 Background to the Washington Program

In 1988 the Washington State Legislature created a Joint Select Committee on Underground Storage Tanks to study and recommend legislation to assist UST owners/operators in complying with the federal financial responsibility requirements. The committee identified many of the problems, previously discussed in this chapter, facing both those who seek financial responsibility and those (such as insurance companies) who provide it. In particular, it recognized the tremendous uncertainties involved in attempting to predict losses, especially for a voluntary state program (which would attract an unknown number of subscribers). "In short, the best forecast of the loss costs of a financial responsibility program amounts to an educated guess."[21] The costs of administering a program were also viewed as uncertain, possibly amounting to as much as 25% of the benefits provided. As well as routine expenses for agents/brokers, underwriters, actuaries, claims adjustors, etc., the committee acknowledged the potential for lawsuits whose loss could involve payments greater than the policy limits. The committee viewed as critical the need to provide adequate funding to cover potential liablity. "Funding must be both sufficient and flexible."[22]

Also viewed as important was the desirability of providing an incentive to reduce risk, which insurance accomplishes through the use of underwriting criteria and variable premiums. State funds, on the other hand, tend to eliminate the incentive by charging fixed fees, and may also displace or eliminate private insurance. A problem with a voluntary state insurance program, however, unless it were to offer some incentive to participate, is that it might attract only the "worst risks" (denying it the opportunity to spread the risks and keep the premiums low). Accordingly, the committee concluded that the state would have to offer prices to "good risks" competitive with those offered privately.

In 1989 the Washington Legislature adopted the Joint Committee's recommendation to create a pollution liability insurance program, subject to further review and approval of the program after 1 year. Following the 1990 review, and having made a few changes, the Legislature authorized full implementation. The program is now in place.

9.7.2 Description of the Washington Program

The underlying objective of the program is to guarantee the availability and affordability of pollution liability insurance for UST owners/operators. The program seeks to achieve *affordability* by having the state sell reinsurance to a pollution liability insurance company "at a price well below the private market price for similar reinsurance. The insurer is required to pass this discount on to owners/operators of underground storage tanks who meet underwriting standards established by the program director."[23] The program seeks to achieve *availability* by "requiring the insurer to accept a greater degree of risk than the insurer would otherwise accept without state involvement."[24]

The program is administered by the Washington State Pollution Liability Agency (PLIA) and relies on three sources of funds: a petroleum products tax (at a rate of 0.5% of the wholesale value of petroleum products on first possession in the state); reinsurance premiums; and interest earned on funds in the PLIA trust account. If the fund grows to $15 million, no more tax is collected until the balance drops below $7.5 million.

Following a solicitation of bids, PLIA has contracted with two private insurance companies (as of May 1991) to provide primary insurance to UST owners/operators.[25] The insurers have agreed to offer policies satisfying federally mandated liability limits to any owner/operator whose tanks:

- Meet the minimum federal requirements
- Are properly registered with the Washington Department of Ecology
- Have appropriate inventory control records
- Meet certain underwriting standards (specified by the insurer) such as a tank tightness test

Program funds may be used to subsidize the costs incurred by an owner/operator in performing the tank or site analyses required by an insurer prior to issuing a policy. The insurer must condition the policy issuance on compliance with laws governing USTs and on the exercise of prudent risk management practices. The policy cannot cover losses resulting from past or existing leaks, but sites containing USTs where a pre-existing leak has been discovered may be eligible for coverage under the program provided that: (1) the owner/operator has a plan for proceeding with corrective action; and (2) until the corrective action has been performed (to the satisfaction of the program director), the burden of proving that a claim is not related to a pre-existing release falls upon the owner/operator. The policy must provide for the insurer to cover defense costs (against a liability claim).

In determining premium levels, an insurer must take into account such factors as location, age, tank design and construction, soil chemistry, distance to surface water, and depth to water table, while generally giving credits for deductibles as well as devices for tank protection, leak detection, and monitoring. If the insurer refuses or cancels coverage, an appeal can be made to the director of the state program; however, the director is likely to proceed cautiously under these circumstances, for he/she is potentially placed in the position of substituting his/her underwriting judgement for that of the company.[26]

For most purposes, the owner/operator deals not with PLIA but with the insurance company, its agent, or its broker, who have responsibility for marketing and sales, tank testing and underwriting requirements, rating and pricing, issuing policies, claims management, and risk management.[27] In the event of a valid claim, once a plan for corrective action has been mutually agreed upon by the insurer and the insured, that action is carried out and the insurer pays the claim in full. The insurer then looks to the owner/operator for reimbursement of the amount of the deductible, and to PLIA for reimbursement of any amount by which the claim exceeds the retention level ($75,000) of the insurance company.[28]

By intent, the program largely builds on standard industry practices. The program is seen as a temporary means to stimulate development of the private insurance market in Washington, and is currently scheduled to terminate in 1995. Its success is not guaranteed, and PLIA could be liquidated in the event of failure. The director is given a high degree of flexibility in running the program, with the expectation that he/she will exercise "sound business judgement" in negotiating with insurers about a variety of insurance and reinsurance variables.[29] The use of sound actuarial principles in designing and managing the program is mandated. The statute provides for a number of reporting/ oversight mechanisms, including the formation of a standing technical advisory committee.

The program offers many incentives to participating insurance companies, including the following:

- Since it receives income from tax revenues, PLIA can offer reinsurance for considerably lower premiums than those that would be charged by a private reinsurer. Thus the insurer retains more of the premiums collected from the tank owners/operators, and also benefits from the investment income earned on these payments.
- Insurers participating in the program are exempted from certain provisions of the Washington insurance code, including a 2% tax (that would otherwise be levied on premiums) and a number of paperwork requirements. Other savings result from a reduction in the required contribution toward financing the state Insurance Commissioner's Office.[30]
- Although policies issued by participating insurers must conform to EPA's financial-responsibility requirements, in other respects they may be designed without regard to any state insurance statute or regulation governing liability policies. Thus the insurers are provided with unusual flexibility in designing, pricing, and marketing their policies under this program.[31]

9.7.3 THE WASHINGTON PROGRAM: DISCUSSION

The Washington program differs from most other state programs in that its major stated goal is to promote the development of a viable *private sector* response to the need for UST financial responsibility. As pointed out by one who designed the program, its success depends on participation by both insurers and tank owners/operators, including the "best" risks as well as the "worst". As previously mentioned, the state has succeeded in establishing contracts with two participating insurance companies (Front Royal and Lloyd's), and the program is underway. According to a PLIA official, about 100 policies have been issued to date, mostly to owners/operators having more than 12 tanks.[32] It is perhaps not surprising that fewer smaller firms have participated so far, in view of the postponed federal financial-responsibility compliance date for these firms.

Whether the program is actually succeeding in its objective of increasing the availability and affordability of insurance for tank owners/operators is not clear. With respect to *availability*, some insurance companies that previously offered policies in Washington (including the companies that currently dominate the UST insurance market nationwide) have declined to participate in the program and are no longer writing tank coverage in the state. A PLIA official, however, claims that more insurers are now trying to enter the program.[33] Front Royal, one of the companies already participating, says that it would have offered coverage in Washington regardless, but that the program has enabled it to significantly broaden its underwriting criteria, thereby extending eligibility to more tank owners/operators.[34] The program also subsidizes the tank tightness test that the company requires of prospective clients (within 2 years of policy issuance).

With regard to *affordability*, an important factor (as previously mentioned) is the level of deductible. Under the state program, the deductible is capped at $50,000, partly because the primary insurer retains only $75,000 of liability (the rest being covered by reinsurance). Front Royal offers deductibles ranging from $10,000 to $50,000, the minimum amount typically increasing with the age of the tank(s) insured. The company normally requires a financial statement before offering a deductible at the higher end of the range; however, most of the 35 or so policies that Front Royal has issued to date carry a deductible of $10,000, for which a financial statement is not required but for which the premium is likely to be relatively high.[35] Policies issued under the program so far by Lloyd's (through Sedgwick James of Spokane, Washington) have also, for the most part, carried small deductibles of $5,000 or $10,000, although an agent has commented that higher deductibles would result in savings in premiums of up to 50%.[36] According to an official of The Planning Corporation (a company that has not sought to participate in Washington's program), "by forcing a low deductible, premium costs become exorbitant and underwriting requirements become draconian. A larger firm would be better off purchasing a policy with a large deductible, thereby substantially reducing premium and underwriting requirements."[37]

An important feature of the Washington program is the sale of reinsurance by the state to participating insurers, at a price lower than market price. It is not clear, however, that this really makes a significant difference to the premium paid by the policy-holder. The main reason is that most losses are expected to fall within the $75,000 amount for which liability is retained by the primary insurer.

Empirical evidence regarding the actual effect of the Washington program on affordability is inconclusive. According to a PLIA official, so far most premiums paid by tank owners/operators to participating insurers within the program have been dramatically less than those previously paid or quoted *for generally similar coverage* outside the program. Figures provided by this official for nine UST policies suggest savings greater than 50%, although in a few cases (according to the official) these savings may partly reflect an upgrading of the tanks being covered.[38] Even with this proviso, however, it is not clear why the savings should be so great, particularly in light of what has been said already about the mostly small deductibles and the relatively minor impact of state-subsidized reinsurance.

A critic of the program (with many years of experience in the UST insurance market) is skeptical about the savings, arguing that there is "no logical reason" why the participating companies should offer them *if the coverage is in fact similar*.[39] A possible explanation, in his opinion, is that the comparison "quotes" provided by the PLIA official might have been given *prior* to the imposition of underwriting requirements (such as a tank-tightness test) whereas the actual premiums would have been established *after* the risks had been better defined. Another explanation has been offered by a representative of Sedgwick James, which issues policies (under the program) for Lloyd's. He says that his company can offer low premiums because it has agreed to handle *in-house* many of the responsibilities normally handled (and charged for) by the underwriters themselves.[40] Since the commission paid by the underwriters has remained the same, it remains to be seen whether this was a wise business decision. Furthermore, without a more detailed investigation, it is impossible to reach definitive conclusions regarding the claimed savings in premiums under the Washington program and the factors that might make these savings possible.

Overall, it is perhaps too early to judge the success of the Washington program. Those who are involved in the program tend to be very positive about it, while others are very critical. As in other states, the greatest challenge is almost certainly posed by small firms, whose ability to obtain insurance at an affordable premium under the Washington program has not, for the most part, yet been demonstrated. A key issue is whether those firms whose tanks do not currently comply with the technical requirements will be financially capable of bringing them into compliance. Recognizing this as a problem, the state now intends to award from the PLIA account up to $15 million in grants to firms located in remote rural areas, with preference given to firms that are the sole source of fuel in these areas. If tanks are brought into technical compliance, however, they are more likely to be eligible for regular insurance coverage, and (according to its critics) the need for the state-subsidized insurance program then becomes increasingly questionable.

NOTES

1. All-Industry Research Advisory Council (1985), p. 2.
2. *Ibid.*
3. FR Nov. 9, 1989.
4. United States General Accounting Office, Underground Storage Tanks: Owners' Ability to Comply With EPA's Financial Responsibility Requirements, GAO/RCED-90-167FS, Washington, D.C., July 1990, 4.
5. Hornbeck, Ethel, S., 1991 Joint Survey of Gasoline Marketer Underground Storage Tank Activity, National Association of Convenience Stores, Petroleum Marketers Association of America, and Society of Independent Gasoline Marketers of America, January 1991, 2.
6. GAO, *op. cit.*, 4.
7. GAO, *op. cit.*, 24.
8. Hornbeck, *op. cit.*, 4.
9. California State Water Control Board, Report on the Availability of Pollution Liability Insurance for Underground Storage Tanks (USTs) Containing Petroleum, LG-98, Sacramento, December 1989, 5.
10. GAO, *op. cit.*, 4.
11. Hornbeck, *op. cit.*, 4.
12. Martin, K., The Planning Corporation, personal communication, April 1991.
13. Anderson, Myra R. and Gulledge, William P., "Do We Need to Upgrade Underground Tank Regulation?" *Risk Management*, August 1989, 36.
14. Anderson and Gulledge, *op. cit.*, 37.
15. Anderson and Gulledge, *op. cit.*, 10.
16. Martin, *op. cit.*
17. Petroleum Marketers Association of America, "Underground Storage Tank Financing Problems: Statement of Need for Guidelines Clarifying Lender Liability," undated memorandum.
18. United States General Accounting Office, Insurers' Pollution Claims Experience, GAO/RCED-91-59, Washington, D.C., 1991.
19. Martin, *op. cit.*
20. Conniff, John S., "Financial Responsibility Assistance for Underground Storage Tanks: Can Washington State Run a Pollution Reinsurance Company?" *Univ. of Puget Sound Law Review*, 14, 1990, 1.
21. Conniff, *op. cit.*, 24.
22. Conniff, *op. cit.*, 29.
23. Conniff, *op. cit.*, 32.
24. Conniff, *op. cit.*, 33.
25. Roney, Richard, Washington Pollution Liability Insurance Agency, personal communication, April 1991.
26. Conniff, *op. cit.*, 49.
27. Washington State Pollution Liability Agency, undated pamphlet.
28. Washington State, *op. cit.*
29. Conniff, *op. cit.*, 35.
30. Conniff, *op. cit.*, 41.
31. Conniff, *op. cit.*, 41.
32. Roney, Richard, Washington Pollution Liability Insurance Agency, personal communication, June 1991.
33. Metelski, John, Front Royal Insurance Co., personal communication, May 1991.
34. Metelski, *op. cit.*
35. Metelski, *op. cit.*
36. McNabb, Bob, Sedgwick James, personal communication, June 1991.
37. Clay, Max, The Planning Corporation, personal communication, May 1991.
38. Roney, *op. cit.*
39. Clay, *op. cit.*
40. McNabb, *op. cit.*

APPENDIX

Examples of UST Insurance Documentation

ENVIRONMENTAL IMPAIRMENT LIABILITY INSURANCE FOR PETROLEUM MARKETERS

ENVIRONMENTAL IMPAIRMENT LIABILITY INSURANCE

> **NOTICE**
>
> PLEASE READ THE ENTIRE CERTIFICATE CAREFULLY. IT PROVIDES LIABILITY COVERAGE ON A CLAIMS MADE BASIS UNDER COVERAGE A, AND ON A NOTICE OF ENVIRONMENTAL IMPAIR~~~ BASIS UNDER COVERAGE B. IT CONTAINS A SINGLE LIMIT OF LIABILITY FOR ALL DAMA~~~ OUT OF THE SAME ENVIRONMENTAL IMPAIRMENT.

SPECIMEN

WE, THE UNDERWRITERS, agree with the NAMED ASSURED, named in the Declarations made a part hereof, in consideration of the payment of the premium and promise to pay the Deductible as described herein, and in reliance upon the statements made in the application forming a part hereof and subject to the limits of liability, exclusions, conditions, and other terms of this Certificate, as follows:

I.   INSURING AGREEMENT

Underwriters hereby agree to pay on behalf of the ASSURED, in excess of the applicable Deductible Amount as set forth in Item 4 of the Declarations, the following amounts:

Coverage A.   All sums that the ASSURED shall be legally obligated to pay for DAMAGES on account of PERSONAL INJURY, PROPERTY DAMAGE, or ENVIRONMENTAL INJURY for CLAIMS first made against the ASSURED and reported to Underwriters during the CERTIFICATE PERIOD, or applicable Extended Reporting Period, arising out of an ENVIRONMENTAL IMPAIRMENT emanating from an INSURED SITE.

Coverage B.   All sums that the ASSURED shall be obligated to pay for DAMAGES on account of POLLUTION CLEAN-UP LIABILITY arising out of an ENVIRONMENTAL IMPAIRMENT emanating from an INSURED SITE which first becomes known to the ASSURED and is reported by or for the ASSURED to Underwriters during the CERTIFICATE PERIOD.

II.   DEFINITIONS

This Certificate is subject to the following definitions:

A.   CLAIM means an oral or written demand made against the ASSURED for DAMAGES, including, without limitation, the service of suit or institution of arbitration proceedings against the ASSURED for DAMAGES.

B.   DAMAGES means:

1. monetary awards or settlements of compensatory damages,

2. reasonable and necessary costs or expenses incurred to remove, treat, neutralize, contain, or clean up any POLLUTANT, including reasonable and necessary costs or expenses incurred in testing for an ENVIRONMENTAL IMPAIRMENT, provided that the testing is conducted in response to objective evidence indicating the presence of an ENVIRONMENTAL IMPAIRMENT or in order to comply with governmental requirements and provided further that the testing leads to the discovery of an ENVIRONMENTAL IMPAIRMENT covered by Underwriters, or

3. DEFENSE EXPENSES;

provided, however, DAMAGES shall not include, and this Certificate shall not pay for the cost of: (a) replacement or repair of storage tanks or other receptacles; (b) replacement or repair of piping, connections and valves in conjunction with (a); (c) excavation or backfilling done in conjunction with (a) or (b); or (d) any testing for an ENVIRONMENTAL IMPAIRMENT not expressly included within the terms of paragraph 2 of this Section B.

C.   DEFENSE EXPENSES means the fees, costs, charges, and expenses incurred by the ASSURED or by Underwriters in the investigation and defense of CLAIMS, excluding all salaries of employees and office expenses of the ASSURED or Underwriters so incurred.

D.   ENVIRONMENTAL IMPAIRMENT means the emission, discharge, disposal, dispersal, release, seepage, or escape of any POLLUTANT into or upon land, the atmosphere, or any watercourse or body of water, provided that such emission, discharge, disposal, dispersal, release, seepage, or escape is neither expected nor intended by the owner or operator of the INSURED SITE or by the ASSURED.

All such emissions, discharges, disposals, dispersals, releases, seepages, or escapes that

i.   are attributable directly or indirectly to the same event(s), circumstance(s), condition(s), or cause(s), or

ii.   emanate from any one INSURED SITE and are the subject of a single program to treat, neutralize, contain, and/or clean up POLLUTANTS in order to comply with applicable legal requirements

shall be treated as one ENVIRONMENTAL IMPAIRMENT, irrespective of the time period or area over which they occur.

E.   ENVIRONMENTAL INJURY means impairment or diminution of or interference with any environmental right or amenity protected by law not included in PERSONAL INJURY or PROPERTY DAMAGE.

F.   ASSURED means the NAMED ASSURED designated as such in the Declarations, and any director, officer, partner, member, employee, or stockholder thereof acting within the scope of his/her duties as such, including heirs, administrators, executors, assigns, and legal representatives of each ASSURED in the event of their death, incapacity, or bankruptcy.

G. INSURED SITE means:

1. each of the sites identified in Item 4 of the Declarations;

2. any residential or farm fuel storage tank, other than a tank at a site identified in Item 4 of the Declarations, that is owned by the NAMED ASSURED, that has a storage capacity of not more than 1100 gallons, and that is used for storing motor fuel for non commercial purposes; and

3. any tank, other than a tank at a site identified in Item 4 of the Declarations, that is owned by the NAMED ASSURED and that is used for storing heating oil for consumptive use on the premises where stored.

H. NAMED ASSURED means the person or entity designated in Item 1 of the Declarations.

I. ORIGINAL CERTIFICATE INCEPTION DATE means the date specified in Item 2 of the Declarations representing the inception date of this Certificate, or, if this Certificate is a renewal policy, the inception date of the first policy issued by the Underwriters to the ASSURED in the continuous and uninterrupted succession of policies that includes this Certificate.

J. PERSONAL INJURY means bodily injury, sickness, disease, mental anguish, shock, or disability sustained by any person, including death resulting therefrom.

K. CERTIFICATE PERIOD means the period set forth in Item 2 of the Declarations, or any shorter period resulting from cancellation.

L. POLLUTANT means any solid, liquid, gaseous, or thermal irritant, contaminant, or toxic or hazardous substance, or any substance that may, does, or is alleged to affect adversely the environment, property, persons, or animals, including smoke, vapors, soot, fumes, acids, alkalis, toxic chemicals, liquids or gases, waste materials, or other irritants or contaminants.

M. POLLUTION CLEAN-UP LIABILITY means:

1. the obligation of the ASSURED to remove, treat, neutralize, contain, or clean up any POLLUTANT in order to comply with any statute, ordinance, rule, regulation, directive, order, or similar legal requirement, in effect at the time the ENVIRONMENTAL IMPAIRMENT giving rise to such obligation is first reported to the Underwriters, of (a) the United States, (b) a State, the District of Columbia, or any political subdivision thereof, or (c) any agency or instrumentality of a governmental body described in (a) or (b), or

2. action by the ASSURED, consented to in advance in writing by the Underwriters, to remove, treat, neutralize, contain, or clean up any POLLUTANT to avert, reduce, or eliminate liability for PERSONAL INJURY, PROPERTY DAMAGE, or ENVIRONMENTAL INJURY that is, or otherwise would have been, covered hereunder.

N. PROPERTY DAMAGE means:

1. physical injury to or destruction of tangible property, including the resulting loss of use thereof, or

2. loss of use of tangible property which has not been physically injured or destroyed, provided such loss of use is caused by an ENVIRONMENTAL IMPAIRMENT.

III. TERRITORY

This Certificate applies only to CLAIMS or POLLUTION CLEAN-UP LIABILITY arising from ENVIRONMENTAL IMPAIRMENTS in the United States, its territories or possessions, or Canada or its territories, but not to any CLAIMS or POLLUTION CLEAN-UP LIABILITY in respect of which suit for DAMAGES is brought other than in the United States, its territories or possessions, or Canada or its territories.

IV. EXCLUSIONS

This insurance does not apply to, and the Underwriters shall have no liability hereunder in respect of, the following:

A. any PERSONAL INJURY, PROPERTY DAMAGE, ENVIRONMENTAL INJURY, or POLLUTION CLEAN-UP LIABILITY arising from:

1. any ENVIRONMENTAL IMPAIRMENT that was known or should have been known to the ASSURED prior to the ORIGINAL CERTIFICATE INCEPTION DATE;

2. any ENVIRONMENTAL IMPAIRMENT with respect to any INSURED SITE added during the CERTIFICATE PERIOD if such ENVIRONMENTAL IMPAIRMENT was known or should have been known to the ASSURED prior to the effective date of coverage for such INSURED SITE;

B. PERSONAL INJURY, PROPERTY DAMAGE, ENVIRONMENTAL INJURY, or POLLUTION CLEAN-UP LIABILITY that arises out of or is directly or indirectly attributable to any failure to comply with any applicable statute, regulation, ordinance, directive, or order relating to the protection of the environment and promulgated by any governmental body, provided that failure to comply is a willful or deliberate act or omission of:

1. the ASSURED;

2. any member, partner, or executive officer thereof, whether or not acting in the scope of their employment; or

3. the owner or operator of an INSURED SITE;

C. any obligation for which the ASSURED or its insurer may be held liable under any Workers' Compensation, unemployment compensation, or disability benefits law or under any similar law;

D. PERSONAL INJURY to any employee of the ASSURED if such injury occurs during and in the course of said employment, or to an obligation of the ASSURED to indemnify another or make contribution to another due to any joint liability because of DAMAGES arising out of such injury;

E. except as provided by COVERAGE B, damage to property:

1. owned by, occupied by, or rented to the ASSURED;

2. used by the ASSURED;

3. in the care, custody, or control of the ASSURED or over which the ASSURED is for any purpose exercising physical control;

F. any liability assumed by the ASSURED under any contract or agreement, oral or written, but this exclusion does not apply to liability that the ASSURED would have in the absence of any such contract or agreement;

G. PROPERTY DAMAGE to goods or products manufactured, sold, handled, or distributed by the ASSURED, or PROPERTY DAMAGE to work performed by or on behalf of the ASSURED;

H. PERSONAL INJURY, PROPERTY DAMAGE, ENVIRONMENTAL INJURY, or POLLUTION CLEAN-UP LIABILITY arising out of an ENVIRONMENTAL IMPAIRMENT emanating from an INSURED SITE that is not reported by or for the ASSURED to Underwriters before the INSURED SITE, or the portion thereof from which the ENVIRONMENTAL IMPAIRMENT emanates, is sold, given away, abandoned, or otherwise transferred by the ASSURED;

I. PERSONAL INJURY, PROPERTY DAMAGE, ENVIRONMENTAL INJURY, or POLLUTION CLEAN-UP LIABILITY arising out of the ownership, maintenance, use, or entrustment to others of any motor vehicle (meaning any land motor vehicle, automobile, trailer, or semi-trailer designed for travel on public roads subject to motor vehicle registration, including any machinery or apparatus attached thereto), aircraft, watercraft, or rolling stock owned or operated by or rented or loaned to any ASSURED. Use includes operation and loading or unloading;

J. PERSONAL INJURY, PROPERTY DAMAGE, ENVIRONMENTAL INJURY, or POLLUTION CLEAN-UP LIABILITY arising out of:

1. any clean-up operation reasonably considered to be routine and normal in connection with the business of the ASSURED;

2. the dumping of any POLLUTANT or radioactive substances in international waters; or

3. the testing, monitoring, cleanup, removal, containment, treatment, detoxification, or neutralization of POLLUTANTS at any waste disposal site or Superfund site;

K. PERSONAL INJURY, PROPERTY DAMAGE, ENVIRONMENTAL INJURY or POLLUTION CLEAN-UP LIABILITY arising out of acid rain;

L. PERSONAL INJURY, PROPERTY DAMAGE, ENVIRONMENTAL INJURY or POLLUTION CLEAN-UP LIABILITY arising out of the ownership or operation of any "offshore facility" as defined in the Outer Continental Shelf Labor Act Amendments of 1978 or the Clean Water Act of 1977, as amended, or any "deepwater port" as defined in the Deepwater Port Act of 1974, as amended;

M. PERSONAL INJURY, PROPERTY DAMAGE, ENVIRONMENTAL INJURY, or POLLUTION CLEAN-UP LIABILITY arising out of the emission, discharge, disposal, dispersal, release, seepage, or escape of drilling fluid, oil, gas, or other fluids or contaminants from any oil, gas, mineral, water, or geothermal well of any nature whatsoever, or from any mine or quarry;

N. 1. PERSONAL INJURY, PROPERTY DAMAGE, ENVIRONMENTAL INJURY, or POLLUTION CLEAN-UP LIABILITY:

a. with respect to which an ASSURED under this Certificate is also an insured under a nuclear energy liability policy issued by Nuclear Energy Liability Insurance Association, Mutual Atomic Energy Liability Underwriters, or Nuclear Insurance Association of Canada, or would be an insured under any such policy but for its termination upon exhaustion of its limits of liability; or

b. resulting from the Hazardous Properties of Nuclear Material and with respect to which (a) any person or organization is required to maintain financial protection pursuant to the Atomic Energy Act of 1954, or any law amendatory thereof, or (b) the ASSURED is, or had this certificate not been issued would be, entitled to indemnity from the United States of America, or any agency thereof, under any agreement entered into by the United States of America, or any agency thereof, with any person or organization;

2. PERSONAL INJURY, PROPERTY DAMAGE, ENVIRONMENTAL INJURY, or POLLUTION CLEAN-UP LIABILITY resulting from the Hazardous Properties of Nuclear Material, if:

a. the Nuclear Material (a) is at any Nuclear Facility owned by, or operated by or on behalf of, an ASSURED or (b) has been discharged or dispersed therefrom;

b. the Nuclear Material is contained in Spent Fuel or Waste at any time possessed, handled, used, processed, stored, transported, or disposed of by or on behalf of an ASSURED; or

c. the injury, damage, or liability arises out of the furnishing by an ASSURED of services, materials, parts, or equipment in connection with the planning, construction, maintenance, operation or use of any Nuclear Facility, but if such facility is located within the United States of America, its territories or possessions or Canada, this exclusion c. applies only to PROPERTY DAMAGE to such Nuclear Facility and any property thereat;

3. as used in this exclusion:

"Hazardous Properties" include radioactive, toxic or explosive properties;

"Nuclear Material" means Source Material, Special Nuclear Material or Byproduct Material;

"Source Material," "Special Nuclear Material," and "Byproduct Material" have the meanings given them in the Atomic Energy Act of 1954 or in any law, amendatory thereof;

"Spent Fuel" means any fuel element or fuel component, solid or liquid, which has been used or exposed to radiation in a Nuclear Reactor;

"Waste" means any waste material (1) containing Byproduct Material and (2) resulting from the operation by any person or organization of any Nuclear Facility included within the definition of Nuclear Facility under paragraph (a) or (b) thereof;

"Nuclear Facility" means

(a) any Nuclear Reactor;

(b) any equipment or device designed or used for (1) separating the isotopes of uranium or plutonium, (2) processing or utilizing Spent Fuel, or (3) handling, processing, or packaging Waste;

(c) any equipment or device used for the processing, fabricating, or alloying of Special Nuclear Material if at any time the total amount of such material in the custody of the ASSURED at the premises where such equipment or device is located consists of or contains more than 25 grams of plutonium or uranium 233, or any combination thereof, or more than 250 grams of uranium 235;

(d) any structure, basin, excavation, premises, or place prepared or used for the storage or disposal of Waste, and includes the site on which any of the foregoing is located, all operations conducted on such site and all premises used for such operations;

"Nuclear Reactor" means any apparatus designed or used to sustain nuclear fission in a self-supporting chain reaction or to contain a critical mass of fissionable material;

PROPERTY DAMAGE includes all forms of radioactive contamination of property;

O. PERSONAL INJURY, PROPERTY DAMAGE, ENVIRONMENTAL INJURY, or POLLUTION CLEAN-UP LIABILITY directly or indirectly occasioned by, happening through or in consequence of war (declared or undeclared), invasion, acts of foreign enemies, hostilities, civil war, rebellion, revolution, insurrection, military or usurped power, or confiscation or nationalization or requisition or destruction of or damage to property by or under the order of any government or public or local authority;

P. PERSONAL INJURY, PROPERTY DAMAGE, ENVIRONMENTAL INJURY, or POLLUTION CLEAN-UP LIABILITY arising out of an ENVIRONMENTAL IMPAIRMENT emanating from an inactive or closed site; provided, however, that this exclusion shall not apply to PERSONAL INJURY, PROPERTY DAMAGE, ENVIRONMENTAL INJURY, or POLLUTION CLEAN-UP LIABILITY arising out of an ENVIRONMENTAL IMPAIRMENT emanating from a site that is only temporarily closed or inactive if:

(1) such site is listed in Item 4 of the Declarations; and

(2) such site continues to be either owned or leased by the NAMED ASSURED; and

(3) the tank system on such temporarily closed or inactive site is operated and maintained in accordance with all applicable rules and regulations, including, but not limited to, 40 C.F.R. § 280.70, as amended;

Q. PERSONAL INJURY, PROPERTY DAMAGE, ENVIRONMENTAL INJURY, or POLLUTION CLEAN-UP LIABILITY included within the "Completed Operations Hazard" or the "Products Hazard," and arising out of an ENVIRONMENTAL IMPAIRMENT that originates away from any INSURED SITE.

For purposes of this exclusion, the term "Completed Operations Hazard" means PERSONAL INJURY, PROPERTY DAMAGE, ENVIRONMENTAL INJURY, or POLLUTION CLEAN-UP LIABILITY arising out of operations or reliance upon a representation or warranty made at any time with respect thereto, but only if the PERSONAL INJURY, PROPERTY DAMAGE, ENVIRONMENTAL INJURY, or POLLUTION CLEAN-UP LIABILITY occurs after such operations have been completed or abandoned and occurs away from premises owned by or rented to the ASSURED. "Operations" include materials, parts, or equipment furnished in connection therewith. Operations shall be deemed completed at the earliest of the following times:

(1) when all operations to be performed by or on behalf of the ASSURED under the contract have been completed; or

(2) when all operations to be performed by or on behalf of the ASSURED at the site of the operations have been completed; or

(3) when the portion of the work out of which the injury or damage arises has been put to its intended use by any person or organization other than another contractor or subcontractor engaged in performing operations for a principal as a part of the same project.

Operations which may require further service or maintenance work, or correction, repair, or replacement because of any defect or deficiency, but which are otherwise complete, shall be deemed completed.

For purposes of this exclusion, the term "Products Hazard" means PERSONAL INJURY, PROPERTY DAMAGE, ENVIRONMENTAL INJURY, or POLLUTION CLEAN-UP LIABILITY arising out of the ASSURED's products or reliance upon a representation or warranty made at any time with respect thereto, but only if the PERSONAL INJURY, PROPERTY DAMAGE, or ENVIRONMENTAL INJURY, or POLLUTION CLEAN-UP LIABILITY occurs away from premises owned by or rented to the ASSURED and after physical possession of such products has been relinquished to others;

R. 1. any governmental, civil, or criminal fines or penalties; or

2. punitive damages of any kind or nature.

No inference shall be made from the express exclusion of liabilities in this Article IV that this Certificate would otherwise cover such liabilities or similar liabilities.

V. EXTENDED REPORTING PERIOD

The NAMED ASSURED shall be entitled to purchase a six-month extension of this coverage from the effective date of cancellation or non-renewal of the policy. Extension of coverage hereunder shall only apply as respects a CLAIM otherwise covered hereunder under COVERAGE A or POLLUTION CLEAN-UP LIABILITY otherwise covered hereunder under COVERAGE B that arises out of an ENVIRONMENTAL IMPAIRMENT commencing prior to the effective date of expiration or non-renewal of this Certificate and which

CLAIM or POLLUTION CLEAN-UP LIABILITY is reported to Underwriters prior to the termination of this Extended Reporting Period. Any extension of coverage hereunder shall not result in the reinstatement of the limits of liability set forth in Item 4 of the Declarations. The premium charged for this Extended Reporting Period shall be as set forth in Section A of Article VIII (Other Conditions) of this Certificate. All other terms, conditions, limitations, and exclusions of this Certificate shall remain unchanged.

VI. LIMITS OF LIABILITY AND DEDUCTIBLE

Irrespective of the period over which the damages, injuries, or liabilities take place, Underwriters' liability under this Certificate shall be limited as follows:

1. In the case of each ENVIRONMENTAL IMPAIRMENT first reported to Underwriters during the CERTIFICATE PERIOD, Underwriters shall only be liable for DAMAGES, in respect of CLAIMS and POLLUTION CLEAN-UP LIABILITY covered under this Certificate or any other certificate issued by Underwriters to the ASSURED, in excess of the Deductible Amount stated in Item 4 of the Declarations, and then, with respect to DAMAGES, exclusive of DEFENSE EXPENSES, up to but not exceeding the Limits of Liability as stated in Item 4 of the Declarations, which Limits of Liability shall be reduced by payments of such Deductible Amount for DAMAGES, except DEFENSE EXPENSES;

2. In the case of each ENVIRONMENTAL IMPAIRMENT first reported to Underwriters during a prior certificate period, Underwriters shall only be liable for DAMAGES, in respect of CLAIMS covered under this Certificate, in excess of the Deductible Amount described in Item 4 of the Declarations, and then, with respect to DAMAGES, exclusive of DEFENSE EXPENSES, up to but not exceeding the Limits of Liability as described in Item 4 of the Declarations, which Limits of Liability shall be reduced by payments of such Deductible Amount for DAMAGES, except DEFENSE EXPENSES.

Subject to the foregoing, Underwriters' Limits of Liability for the CERTIFICATE PERIOD and inclusive of the Extended Reporting Period, if applicable, in the aggregate for all DAMAGES, exclusive of DEFENSE EXPENSES, in respect of CLAIMS (covered hereunder under COVERAGE A) and all POLLUTION CLEAN-UP LIABILITY (covered hereunder under COVERAGE B) shall not exceed the Limits of Liability as stated in Item 4 of the Declarations.

In no event shall the inclusion of POLLUTION CLEAN-UP LIABILITY as provided in COVERAGE B increase Underwriters' Limits of Liability as set forth in Item 4 of the Declarations, nor shall the inclusion of more than one ASSURED serve to increase Underwriters' Limits of Liability as stated therein.

The Deductible Amount applies to the payment of DAMAGES, including DEFENSE EXPENSES. Underwriters shall, to the extent necessary for this Certificate to comply with applicable financial responsibility regulations promulgated by the Environmental Protection Agency pursuant to 42 U.S.C. § 6991 b(d), as amended, pay the Deductible Amount. If Underwriters pay part or all of the Deductible Amount for any reason, upon notification of the action taken, the NAMED ASSURED shall promptly reimburse Underwriters for such part of the Deductible Amount as has been paid by Underwriters.

VII. NOTICE OF ENVIRONMENTAL IMPAIRMENT; CLAIM OR SUIT PROVISIONS

A. If the ASSURED becomes aware of an ENVIRONMENTAL IMPAIRMENT which could involve this Certificate, Underwriters shall be given:

1. immediate written or oral notice containing particulars sufficient to identify the ASSURED;

2. reasonably obtainable information with respect to the time, place, and circumstances thereof; and

3. the names and addresses of any injured, and of any available witnesses.

The above information shall be given by or for the ASSURED to Underwriters through its representative at the following address: **THE PLANNING CORPORATION, 11347 SUNSET HILLS ROAD, RESTON, VIRGINIA 22090.** In the event of oral notice, the ASSURED agrees to furnish a written report as soon as practicable. Failure to provide notice as required herein shall result in a forfeiture of any rights to coverage in respect of such ENVIRONMENTAL IMPAIRMENT under this Certificate or any other policy issued by Underwriters to the ASSURED. Subject to the ASSURED's obligation under this Certificate to take promptly all reasonable steps necessary to prevent injury or damage from arising, no costs, charges, or expenses shall be incurred without Underwriters' consent, which shall not be unreasonably withheld.

B. A CLAIM is "first made" within the meaning of this Certificate when the ASSURED first becomes aware of the CLAIM.

C. In the event of a CLAIM or suit, immediate written or oral notice containing particulars sufficient to identify the ASSURED, and reasonably obtainable information with respect to the time, place, and circumstances thereof, and the names and addresses of the injured and of available witnesses, shall be given by or for the ASSURED to Underwriters through its representative at the following address: **THE PLANNING CORPORATION, 11347 SUNSET HILLS ROAD, RESTON, VIRGINIA 22090.** In the event of oral notice, the ASSURED agrees to furnish a written report as soon as possible. Failure to provide notice as required herein shall result in a forfeiture of any rights to coverage in respect of such CLAIM under this Certificate or any other policy issued by Underwriters to the ASSURED.

D. If CLAIM is made or suit is brought against the ASSURED, the ASSURED shall immediately forward every demand, notice, summons, or other process received by him or his representative to Underwriters through its claims administrator at the following address: **WILSON, ELSER, MOSKOWITZ, EDELMAN & DICKER, 420 LEXINGTON AVENUE, NEW YORK, NEW YORK 10170.**

E. Underwriters shall have the right and duty to defend any claim or suit against the ASSURED seeking DAMAGES to which COVERAGE A applies. The ASSURED, on demand of Underwriters, shall promptly reimburse Underwriters for any DEFENSE EXPENSES incurred by Underwriters falling within the ASSURED's Deductible Amount. Underwriters may, at its discretion, investigate any ENVIRONMENTAL IMPAIRMENT and settle any claim or suit that may result without the consent of the ASSURED. Underwriters' right and duty to defend end and Underwriters shall not be obligated to continue to defend any claim or suit, when the applicable limit of Underwriters' liability has been exhausted.

F.　The ASSURED and any of its members, partners, officers, directors, administrators, stockholders, and employees that Underwriters deem necessary agree to cooperate with Underwriters in the investigation, settlement, or defense of any claim or suit, and, upon Underwriters' request, assist in (a) making settlements, (b) the conduct of suits or proceedings, and (c) enforcing any right against any person or organization that may be liable to the ASSURED because of injury or damage with respect to which insurance is afforded under this Certificate, all without charge to Underwriters. The ASSURED, and any of its members, partners, officers, directors, administrators, stockholders, and employees that Underwriters deem necessary, shall attend hearings and trials and assist in securing and giving evidence and obtaining the attendance of witnesses, all without charge to Underwriters. With respect to any claim or suit, the ASSURED shall not, except at the ASSURED's own cost, voluntarily make any settlement or payment, assume any obligation, admit any liability, or incur any expense, without Underwriters' written consent.

G.　Underwriters shall have the right, but not the duty, to participate at its expense in any governmental proceeding that seeks to impose legal obligations in respect of an ENVIRONMENTAL IMPAIRMENT covered hereunder.

VIII. OTHER CONDITIONS

A.　PREMIUMS—The premium for the CERTIFICATE PERIOD shall be as set forth in Item 4 of the Declarations.

An additional premium of up to 50% of the sum of the premium shown in Item 4 of the Declarations and in any endorsements to this Certificate shall be paid to the Underwriters upon cancellation or non-renewal of this Certificate to purchase the six-month extension of coverage (other than any return premium shown in a Certificate termination endorsement) available under Article V (Extended Reporting Period) of this Certificate. Payment of the required Extended Reporting Period premium shall be made not later than 10 days prior to the effective date of cancellation or non-renewal of this Certificate.

B.　INSPECTION AND AUDIT—Underwriters shall be permitted but not obligated to inspect the ASSURED's property and operations at any time. Neither Underwriters' right to make inspection nor the making thereof nor any report thereon shall constitute an undertaking, on behalf of or for the benefit of the ASSURED or others, to determine or warrant whether such property or operations are safe or healthful, or are in compliance with any law, rule, or regulation.
Underwriters may examine and audit the ASSURED's books and records at any time during the CERTIFICATE PERIOD and extension thereof and within three years after the final termination of this Certificate, as far as the aforesaid books and records relate to the subject matter of this insurance.

C.　PREVENTION OF LOSSES—In the event of any ENVIRONMENTAL IMPAIRMENT or knowledge of any circumstances, not previously disclosed, that might reasonably be expected to result in an ENVIRONMENTAL IMPAIRMENT, the ASSURED shall promptly take all reasonable steps to prevent injury or damage from arising out of the impairment or circumstances and shall give notice of such circumstances or preventative measures to the Underwriters through its representative at the following address:
THE PLANNING CORPORATION, 11347 SUNSET HILLS ROAD, RESTON, VIRGINIA 22090.

D.　PROMPT COMPLIANCE WITH CLEAN-UP OBLIGATION – In the event of any POLLUTION CLEAN-UP LIABILITY that involves the legal requirement of the ASSURED to remove, treat, neutralize, contain, or clean up any POLLUTANT in order to comply with any statute, ordinance, rule, regulation, directive, order, or similar legal requirement, the ASSURED shall, subject to first securing Underwriters' consent to the extent required under this Certificate, promptly undertake all reasonable steps necessary to satisfy its obligation to comply with such legal requirement. Failure to take such prompt action shall result in forfeiture of any rights to coverage under this Certificate for such POLLUTION CLEAN-UP LIABILITY.

E.　CANCELLATION – This Certificate may be cancelled by the NAMED ASSURED by surrender thereof to Underwriters or any of its authorized representatives, or by mailing to Underwriters written notice stating when thereafter the cancellation shall be effective. This Certificate may be cancelled by the Underwriters by mailing, by certified mail return receipt requested, to the NAMED ASSURED at the address shown in Item 1 of the Declarations, written notice stating when such cancellation shall be effective, not less than 60 days (10 days for non-payment of premium or misrepresentation) thereafter. The mailing of notice as aforesaid shall be sufficient proof of notice. The time of surrender, or the effective date and hour of cancellation, stated in the notice shall become the end of the CERTIFICATE PERIOD. Delivery of such written notice either by the NAMED ASSURED or by Underwriters shall be equivalent to mailing.

If the NAMED ASSURED cancels, earned premium shall be computed in accordance with the customary short-rate procedure. If Underwriters cancel, earned premium shall be computed pro rata. Premium adjustment may be made either at the time cancellation is effected, or as soon as practicable after cancellation becomes effective, but payment or tender of unearned premium is not a condition of cancellation.

F.　DECLARATIONS – By acceptance of this Certificate, the ASSURED agrees that the statements in the Declarations and application are the ASSURED's agreements and representations and that this Certificate is issued in reliance upon the truth of such representations and that this Certificate embodies all agreements existing between the ASSURED and Underwriters or any of its representatives relating to this insurance.

G.　ACTION AGAINST COMPANY – No action shall lie against Underwriters unless, as a condition precedent thereto, there shall have been full compliance with all of the terms of this Certificate, nor until the amount of the ASSURED's obligation to pay shall have been finally determined either by judgment against the ASSURED after actual trial or by written agreement of the ASSURED, the claimant, and Underwriters. It is agreed that in the event of the failure of Underwriters to pay an amount claimed to be due hereunder, Underwriters, at the request of the NAMED ASSURED, shall submit to the jurisdiction of any court of competent jurisdiction within the United States and shall comply with all requirements necessary to give such court jurisdiction, and all matters arising hereunder shall be determined in accordance with the law and practice of such court. It is further agreed that service of process in such suit may be made upon **WILSON, ELSER, MOSKOWITZ, EDELMAN & DICKER, 420 LEXINGTON AVENUE, NEW YORK, NEW YORK 10170** and that in any suit instituted against it upon this contract, Underwriters shall abide by the final decision of such court or of any appellate court in the event of an appeal.

Any person or organization or the legal representative thereof who has secured a judgment or written agreement in accordance with the foregoing shall thereafter be entitled to recover under this Certificate to the extent of the insurance afforded by this Certificate. No person or organization shall have any right under this Certificate to join Underwriters as a party to any action against the ASSURED to determine the ASSURED's liability, nor shall Underwriters be impleaded by the ASSURED or his legal representative. Bankruptcy or insolvency of the ASSURED or the ASSURED's estate shall not relieve Underwriters of any of its obligations hereunder.

Further, pursuant to any statute of any state, territory or district of the United States which makes provision therefore, Underwriters hereon hereby designate the Superintendent, Commissioner or Director of Insurance or other officer specified for that purpose in the statute, or his successor or successors in office, as their true and lawful attorney upon whom may be served any lawful process in any action, suit or proceeding instituted by or on behalf of the ASSURED or any beneficiary hereunder arising out of this contract of insurance, and hereby designate the above-named as the person to whom the said officer is authorized to mail such process or a true copy thereof.

H.   ASSIGNMENT – Assignment of interest under this Certificate shall not bind Underwriters until its consent is endorsed thereon.

I.   SUBROGATION – In the event of any payment under this Certificate, Underwriters shall be subrogated to all the ASSURED's rights of recovery therefor against any person or organization, and the ASSURED shall execute and deliver instruments and papers and do whatever else is necessary to secure such rights. The ASSURED shall do nothing after loss to prejudice such rights.

J.   OTHER RECOVERIES – In the event an ASSURED recovers, from any governmental fund available to the ASSURED or otherwise, any DAMAGES covered under this Certificate, Underwriters shall be entitled to recover from the ASSURED the portion of such recovery that is equal to the amount of such DAMAGES paid by Underwriters in excess of any Deductible Amount paid by the ASSURED.

K.   CHANGES - Notice to any agent or knowledge possessed by any agent or by any other person shall not effect a waiver or a change in any part of this Certificate or estop Underwriters from asserting any right under the terms of this Certificate nor shall the terms of this Certificate be waived or changed, except by endorsement issued to form a part of this Certificate.

L.   OTHER INSURANCE - This insurance shall be excess of any other valid and collectible insurance, whether such other insurance is stated to be primary, contributory, excess, contingent, or otherwise, unless such other insurance is written only as specific excess insurance over the Limits of Liability provided in this Certificate. When this insurance is excess, Underwriters will have no duty to defend any claim or "suit" that any other insurer has a duty to defend. If no other insurer defends, Underwriters will undertake to do so, but Underwriters will be entitled to the ASSURED's rights against all other insurers.

M.   SOLE AGENT - The NAMED ASSURED shall act on behalf of all ASSUREDS for payment or return premium, receipt and acceptance of any endorsement issued to form a part of this Certificate, giving and receiving notice of cancellation or non-renewal, and the exercise of the rights provided in the Extended Reporting Period clause.

NEW SITES ENDORSEMENT

It is hereby understood and agreed that coverage under this Certificate is limited to sites and storage tanks included within the meaning of the term INSURED SITE as defined in the Certificate and that this Certificate does **not** automatically extend coverage to newly acquired or activated sites.

The ASSURED may apply to Underwriters for coverage for newly acquired or activated sites by providing Underwriters with a written request for coverage describing the sites and storage tanks for which coverage is desired. Underwriters reserve the right to vary the terms and conditions of, or charge an additional premium for, such coverage, or to decline to cover any or all of the newly acquired or activated sites or storage tanks for which coverage is applied.

If Underwriters approve coverage for any newly acquired or activated site, such coverage shall take effect as of the date specified in the endorsement issued by Underwriters to the ASSURED pursuant to which such coverage is provided.

All other terms and conditions remain unchanged.

LLOYD'S OF LONDON
(THROUGH THE PLANNING CORPORATION)

 # Lloyd's, London

This Insurance is effected with certain Underwriters at Lloyd's, London (not incorporated).

This Certificate is issued in accordance with the limited authorization granted to Whistondale & Partners plc by certain Underwriters at Lloyd's, London whose names and the proportions underwritten by them can be ascertained from the office of said Correspondent (such Underwriters being hereinafter called "Underwriters") and in consideration of the premium specified herein, Underwriters do hereby bind themselves each for his own part, and not one for another, their heirs, executors and administrators.

The Assured is requested to read this certificate, and if not correct, return it immediately to the Correspondent for appropriate alteration, via the Planning Corporation, USA.

In the event of a claim under this certificate, please notify the following, The Planning Corporation, 11347 Sunset Hills Road, Reston, Virginia 22090, USA.

Tel.:   0101 703 481-0200
Fax: 0101 703 481-5667

SLC-3 (USA) 1/89 REV.
EASON PRINTING CO., CHICAGO

ENVIRONMENTAL IMPAIRMENT
LIABILITY INSURANCE FOR
PETROLEUM MARKETERS

Underwriter

Coverage is provided by certain underwriters at Lloyds, London.

Limits

Limits currently available are $1,000,000 per environmental impairment with a $2,000,000 annual aggregate.

Deductibles are available from $50,000 to $250,000.

Coverage can be tailored to "wrap-around" state tank funds.

Policy Covers

Claims made against an insured and reported to the underwriter while coverage is in effect for losses which first become known (no limiting retroactive date) after the first policy is issued and which arise out of:

- personal injury
- property damage (including pollution clean-up), or
- environmental impairment of air, land or water.

Pollution clean-up (both on premises and off) required by law which arise out of a pollution incident which becomes known to the insured and is reported to the underwriter during the policy period.

Both gradual and sudden and accidental pollution losses.

Defense costs.

POLICY IS DESIGNED TO MEET FEDERAL FINANCIAL RESPONSIBILITY REGULATIONS

SPECIAL DISCOUNTS AVAILABLE TO MEMBERS OF SPONSORING ASSOCIATIONS

For further information contact: The Planning Corporation
                                          11347 Sunset Hills Road
                                          Reston, VA 22090
                                          (703) 481-0200

CERTIFICATE PROVISIONS

1. **Signature Required.** This certificate shall not be valid unless signed by the Correspondent on the attached Declaration Page.

2. **Correspondent Not Insurer.** The Correspondent is not an Insurer hereunder and neither is nor shall be liable for any loss or claim whatsoever. The Insurers hereunder are those individual Underwriters at Lloyd's, London whose names can be ascertained as hereinbefore set forth.

3. **Cancellation.** If this certificate provides for cancellation and this certificate is cancelled after the inception date, earned premium must be paid for the time the insurance has been in force.

4. **Service of Suit.** It is agreed that in the event of the failure of the Underwriters hereon to pay any amount claimed to be due hereunder, the Underwriters hereon, at the request of the Insured, will submit to the jurisdiction of a Court of competent jurisdiction within the United States. Nothing in this Clause constitutes or should be understood to constitute a waiver of Underwriters' rights to commence an action in any Court of competent jurisdiction in the United States, to remove an action to a United States District Court, or to seek a transfer of a case to another Court as permitted by the laws of the United States or of any State in the United States. It is further agreed that service of process in such suit may be made upon the firm or person named in item 6 of the attached Declaration Page, and that in any suit instituted against any one of them upon this contract. Underwriters will abide by the final decision of such Court or of any Appellate Court in the event of an appeal.

The above-named are authorized and directed to accept service of process on behalf of Underwriters in any such suit and/or upon the request of the Insured to give a written undertaking to the Insured that they will enter a general appearance upon Underwriters' behalf in the event such a suit shall be instituted.

Further, pursuant to any statute of any state, territory, or district of the United States which makes provision therefore, Underwriters hereon hereby designate the Superintendent, Commissioner or Director of Insurance or other officer specified for that purpose in the statute, or his successor or successors in office, as their true and lawful attorney upon whom may be served any lawful process in any action, suit or proceeding instituted by or on behalf of the Insured or any beneficiary hereunder arising out of this contract of insurance, and hereby designate the above-named as the person to whom the said officer is authorized to mail such process or a true copy thereof.

5. **Assignment.** This certificate shall not be assigned either in whole or in part without the written consent of the Correspondent endorsed hereon.

6. **Attached Conditions Incorporated.** This certificate is made and accepted subject to all the provisions, conditions and warranties set forth herein, attached, or endorsed, all of which are to be considered as incorporated herein.

7. **Short Rate Cancellation.** If the attached provisions provide for cancellation, the table below will be used to calculate the short rate proportion of the premium when applicable under the terms of cancellation.

Short Rate Cancellation Table For Term of One Year

Days Insurance in Force	Per Cent of One Year Premium	Days Insurance in Force	Per Cent of One Year Premium	Days Insurance in Force	Per Cent of One Year Premium	Days Insurance in Force	Per Cent of One Year Premium
1	5%	66 - 69	29%	154 - 156	53%	256 - 260	77%
2	6	70 - 73	30	157 - 160	54	261 - 264	78
3 - 4	7	74 - 76	31	161 - 164	55	265 - 269	79
5 - 6	8	77 - 80	32	165 - 167	56	270 - 273(9 mos.)	80
7 - 8	9	81 - 83	33	168 - 171	57	274 - 278	81
9 - 10	10	84 - 87	34	172 - 175	58	279 - 282	82
11 - 12	11	88 - 91(3 mos.)	35	176 - 178	59	283 - 287	83
13 - 14	12	92 - 94	36	179 - 182(6 mos.)	60	288 - 291	84
15 - 16	13	95 - 98	37	183 - 187	61	292 - 296	85
17 - 18	14	99 - 102	38	188 - 191	62	297 - 301	86
19 - 20	15	103 - 105	39	192 - 196	63	302 - 305(10 mos.)	87
21 - 22	16	106 - 109	40	197 - 200	64	306 - 310	88
23 - 25	17	110 - 113	41	201 - 205	65	311 - 314	89
26 - 29	18	114 - 116	42	206 - 209	66	315 - 319	90
30 - 32(1 mo.)	19	117 - 120	43	210 - 214(7 mos.)	67	320 - 323	91
33 - 36	20	121 - 124(4 mos.)	44	215 - 218	68	324 - 328	92
37 - 40	21	125 - 127	45	219 - 223	69	329 - 332	93
41 - 43	22	128 - 131	46	224 - 228	70	333 - 337(11 mos.)	94
44 - 47	23	132 - 135	47	229 - 232	71	338 - 342	95
48 - 51	24	136 - 138	48	233 - 237	72	343 - 346	96
52 - 54	25	139 - 142	49	238 - 241	73	347 - 351	97
55 - 58	26	143 - 146	50	242 - 246(8 mos.)	74	352 - 355	98
59 - 62(2 mos.)	27	147 - 149	51	247 - 250	75	356 - 360	99
63 - 65	28	150 - 153(5 mos.)	52	251 - 255	76	361 - 365(12 mos.)	100

Rules applicable to insurance with terms less than or more than one year:

A. If insurance has been in force for one year or less, apply the short rate table for annual insurance to the full annual premium determined as for insurance written for a term of one year.

B. If insurance has been in force for more than one year:

1. Determine full annual premium as for insurance written for a term of one year.

2. Deduct such premium from the full insurance premium, and on the remainder calculate the pro rata earned premium on the basis of the ratio of the length of time beyond one year the insurance has been in force to the length of time beyond one year for which the policy was originally written.

3. Add premium produced in accordance with items (1) and (2) to obtain earned premium during full period insurance has been in force.

AGRICULTURAL EXCESS & SURPLUS INSURANCE COMPANY (AESIC) (THROUGH FRANK B. HALL OF MISSOURI, INC.)

AESIC
Agricultural Excess & Surplus Insurance Company

Underground Storage Tank Program Questions and Answers

Q.   What coverages are provided under the policy?
A.   1.   Policy is designed to meet all the EPA financial
          responsibility requirements for Underground Storage
          Tanks.
     2.   Pollution Liability including government mandated
          clean up costs (on and off site).
     3.   Third Party Bodily Injury and Property Damage.

Q.   Is the policy claims made or occurrence?
A.   The policy is a claims made policy.

Q.   Are defense costs provided?
A.   Yes. A $100,000 defense cost limit (higher limits are
     optional) is provided in addition to the policy limit.

Q.   Are "prior acts" covered by the policy?
A.   No. Pollution incidents that the insured knows about
     prior to the Retroactive date of the policy are not
     covered.

Q.   Does the policy provide an Extended Reporting Period?
A.   Yes. A 12 month Extended Reporting Period is standard.
     This exceeds the EPA requirement for 6 months.

Q.   What limits does the policy provide?
A.   The policy will provide the limits of insurance needed
     to satisfy the EPA requirements, $1,000,000 per
     pollution incident, and 2,000,000 annual aggregate will
     be most common. The limits apply separately to each
     insured location. Higher limits are available.

Q.   What is the Minimum Annual Premium?
A. $3,500 policy writing minimum.

Environmental
Risk Division

Q. What deductibles are available?
A. $5,000 per pollution incident is the minimum. Higher deductibles are available depending on the insured's financial condition.

Q. What type of applicants will be considered for coverage by the program?
A. Any owner and/or operator of one or more petroleum USTs. This includes both petroleum marketers (gas stations, C-stores, etc.) as well as non-marketers.

Q. What type of tanks are eligible for coverage?
A. Initially coverage will be limited to petroleum underground tanks systems only (all gasoline blends, waste oil, diesel, fuel oil, kerosene, aviation gasoline, Jet A and motor oil). Above ground tank will be considered and can be added by endorsement to the UST policy.

Q. Are there any age restrictions on the tanks that can be covered?
A. No. We will consider any tank regardless of age or construction, but older unprotected tanks will need inspections and confirmation that they are not leaking and the site is not currently contaminated.

Q. Will AESIC provide coverage for chemical tanks?
A. No. Coverage is applicable to petroleum tanks only.

Q. Who pays for the inspections?
A. The applicant/insured will pay for all tests that are part of the underwriting requirements.

Q. How are testing requirements determined?
A. AESIC underwriters in Cincinnati will determine if soil sampling or precision leak testing is necessary based on the UST system, construction, age, spill/overfill prevention and leak detection system(s) on a site specific basis.

Environmental
Risk Division

Q.   How does the program address State Trust Funds?
A.   The coverage will not be used to wrap around State
      funds. The policy is designed to provide clean up and
      third party liability from the first dollar. If
      coverage is also available from the State Trust Fund,
      AESIC will seek reimbursement from the Fund.

Q.   If a State Trust Fund satisfied the EPA financial
      requirements, why would an insured buy the policy?
A.   The answer varies due to differences in State Trust
      Funds. However, certain characteristics are peculiar
      to most of the funds.
      1. None of the states guarantee the payment of
          claims.
      2. Very few states offer the full EPA mandated
          coverages.
      3. Some states only pay a loss over a large deductible
          or SIR amount.
      4. Most states are very slow in paying losses.
      5. Most states are drastically underfunded to pay
          claims.
      6. Most states lack experienced UST or insurance
          personnel to administer the program.
      7. State Funds do not provide coverage for defense
          costs.
      8. Insured must pay all costs first then reimbursement
          can be sought.

Q.   Will AESIC provide coverage excess of another company
      or self-insured retention?
A.   No. AESIC will pay all claims from dollar one, seek
      reimbursement from the insured for the applicable
      deductible amount and seek reimbursement from any
      applicable state UST fund.

Q.   If the insured desired to change coverage to a
      different insurance company in a future year after
      renewal, what happens to the extended reporting period
      under AESIC policy?
A.   The extended reporting period is automatically
      activated for 12 months. During this time polluting
      incidents during the policy period can be reported to
      AESIC. Once a pollution incident is reported to the
      company, all claims and clean up costs are covered
      (unless the limits are exhausted) whenever they are
      incurred.

Q.   How are claims to be handled?
A.   There is a 24 hour "800" number for AESIC UST claims
      reporting. In event of a known or suspected pollution
      incident the insured will notify AESIC immediately.
      Upon notification an engineering emergency response
      team will be dispatched to the site.

Environmental *Frank Hall* *Risk Division*

AESIC
<u>POLICY HIGHLIGHTS</u>

* Three separate insuring agreements:

    1. Third party bodily injury and property damage liability
    2. Government mandated clean-up cost liability
    3. Defense costs

* Claims-made.

* Site specific.

* Retroactive date (no coverage for known incidents
  commencing prior to the retroactive date), retroactive
  cover may be provided when replacing other coverage.

* No coverage for property damage, environmental damage or
  bodily injury emanating from a waste facility.

* No coverage for autos.

* Coverage for loading/unloading of UST's.

* No coverage for the costs of testing, monitoring and
  determining the source and extent of contamination.

* No coverage for punitive damages.

* Coverage is limited to UST systems only. However, above
  ground and bulk facilities will be considered.

* Limits apply <u>separately</u> to each insured <u>site.</u>

* Defense costs are provided under a <u>separate</u> per incident
  limit.

* The deductible reduces the limit of insurance.

* Claims, suits and defense costs apply to the deductible.

* AESIC will pay deductible and seek reimbursement from the
  insured.

* Underground Tank Data Sheets are part of the policy.

* Premium is a flat charge which must be received in full at
  policy inception.

* An "800" number has been set up to report claims/incidents
  immediately to AESIC.

* 60 day notice of non-renewal and cancellation.

* 12 month automatic extended reporting period if policy is
  canceled or non-renewed either by AESIC <u>or</u> by the insured.

Environmental
Risk Division

NATIONAL UNION FIRE INSURANCE COMPANY OF PITTSBURG (THROUGH SEDGWICK JAMES OF PENNSYLVANIA, INC.)

Sedgwick James

Sedgwick James of Pennsylvania, Inc.
PO Box 1675, Harrisburg, Pennsylvania 17105
Telephone (717) 763-7261. Telex 842-382. Facsimile (717) 763-8575

ENVIROGUARD

Thank you for your inquiry regarding the EnviroGuard Program. EnviroGuard provides a "stand-alone" Pollution Legal Liability Policy for operators of underground storage tanks. Highlights of the program include:

*Limits	$500,000 per occurrence/$1,000,000 aggregate up to $2,000,000 per occurrence/$5,000,000 aggregate
*Policy Form	Claims Made (copy enclosed)
*Retroactive Date	Policy inception (any mid-term site additions will have a retroactive date corresponding the the addition date)
*Carrier	A+ rated carrier admitted in all stat.
*Minimum Premium	$2,000
*Minimum Deductible	$10,000 (deductibles up to $250,000 available)
*Coverage	Off-site clean-up Third-party liability (bodily injury and property damage)
*Additional Insureds	Owner may be added to operator's policy for 10% additional premium
*Capital Contribution	None. This is a fully-insured program, not a risk retention group
*Size of Account Requirements	None, any size of account is eligible

The vast majority of the states in the country have established, or are in the process of implementing, funds for the purpose of providing on-site clean-up coverage subject to minimal retentions ranging from $10,000 to $100,00. For this reason the EnviroGuard policy currently excludes on-site clean-up in most instances, however, this coverage <u>is available</u> on an account-by-account basis subject to certain underwriting criteria.

Chapter 10

The Use of State Funds as a Financial Responsibility Mechanism

Paul Thompson

CONTENTS

0-87371-402-4/93/$0.00+$.50
© 1993 by Lewis Publishers

The Use of State Funds as a Financial Responsibility Mechanism

10.1 INTRODUCTION

State funds are the financial responsibility mechanism on which most underground storage tanks (UST) owners/operators must depend to comply with EPA's financial responsibility regulations. As indicated in the previous chapters, insurance still is not widely available and the other financial responsibility mechanisms allowed by EPA generally require UST owners/operators to have generated required levels of financial responsibility themselves. Not surprisingly, the vast majority of firms in the regulated community do not generate sufficient profits to put aside between $1 million and $2 million for a financial responsibility mechanism dedicated only to UST corrective action and third party compensation. Thus, most UST owners/operators currently are looking for public assistance, in the form of a state financial responsibility fund, to provide these funds.

On the surface, the use of a state fund to demonstrate financial responsibility for UST leaks and comply with EPA's regulations seems like a fairly straightforward process and an attractive option for many UST owners/operators. The state simply establishes a fund that is acceptable to EPA, and the UST owner/operator is then covered by a financial responsibility mechanism that meets the federal requirements. In practice, however, the use of a state fund to comply with EPA financial responsibility regulations is not so simple. State funds vary widely and many still impose some financial responsibility requirements on UST owners/operators.

Some funds may provide only partial coverage for the UST owner/operator. Others may allow UST owners/operators to comply with the UST financial responsibility regulations but still require the owner/operator to make large expenditures if an UST leak occurs. Still others may provide certain classes or categories of UST owners/operators with a sound financial responsibility mechanism that not only complies with the EPA regulations, but also covers the cost of an UST leak without requiring reimbursement by the responsible UST-owning firm.

Perhaps most importantly, a state financial assurance fund can be successful only under two scenarios. First, the fund must be sustained over the longterm as a viable mechanism for UST owners/operators to use in complying with EPA's financial responsibility regulations. As discussed briefly in Chapter 7, however, the high costs in public dollars and departure from the "polluter pays principle" could make it difficult for funds to provide the levels of cleanup and third party compensation needed for the existing UST population. Generally, state assumption of large private liabilities (like the billions of dollars needed to clean up existing leaking USTs) is neither politically popular nor economically efficient.

Second, state financial assurance funds could serve as a temporary "stop gap" measure to be used until the UST insurance industry grows to provide a lasting, affordable financial responsibility mechanism for large segments of the regulated community. But as discussed in Chapters 7 and 9, state funds could actually *decrease* the availability of UST insurance. This is evidenced by surveys showing that UST owners/operators are forgoing insurance in favor of state financial responsibility funds, which are less expensive for the owner/operator but more expensive for the public.

Presumably, after all USTs have been upgraded, all existing releases have been cleaned up, and all third-party damages have been paid, the liability risks facing state funds or private UST insurance will be much lower than either currently face. Most current UST leaks will be discovered in the next 8 years (as EPA's UST technical regulations are phased-in), making this the most difficult period for UST owners/operators to obtain an adequate financial responsibility mechanism. To understand how a state fund can assist them, not only in complying with EPA's financial responsibility regulations, but in providing a sound financial resource in the case of an UST release, every owner/operator needs to answer the following two questions:

1. Will a state fund provide the coverage I need to comply with UST financial responsibility regulations until I can qualify for, afford, and obtain private UST insurance?
2. Will the state fund provide the coverage I need in the event that a leak is discovered in one or more of my USTs?

This chapter helps to answer these questions by explaining the types of state funds that EPA will approve as a financial responsibility mechanism, some of the problems that certain types of funds may cause UST owners/operators, and the progress that states are making in establishing state funds.

10.2 EPA GUIDANCE FOR APPROVING STATE FUNDS

State financial assurance funds differ dramatically in the amount and scope of coverage provided and many are being updated as new information is collected on the UST insurance market and the feasibility of current funding approaches. Some funds may provide all of the coverage needed for all UST-owning firms to comply with EPA's financial responsibility regulations while others may provide only part of this coverage for a limited category of UST owners/operators. Different types of state funds have different advantages and disadvantages — both for the UST owner/operator and the public.

EPA has issued guidance to states to indicate how the agency will determine whether to approve a proposed state financial assurance fund.* The guidance provides states with significant flexibility in the administration and financing of these funds. In addition to explaining the different types of state funds that will be used to demonstrate financial responsibility, this guidance reveals why existing funds vary so much from one another in the types and levels of coverage provided.

Understanding how EPA expects state funds to be structured can help UST owners/operators determine whether a particular state fund will meet their financial responsibility needs. UST owners/operators should recognize that EPA approval of a state fund does not guarantee that the fund will provide the full financial responsibility coverage required by EPA or even conform to the guidelines used by the agency to approve the state fund. The guidance states that state funds can be used to provide only partial financial responsibility coverage. In addition, the EPA-approval process is largely based on the subjective judgment of EPA officials and does not strictly apply objective review criteria to state fund proposals.

Thus, UST owners/operators cannot make any assumptions about a particular state fund — or fund proposal — before carefully considering the individual fund in question. In other words, even if a state fund has been — or will be — approved by EPA as an acceptable financial responsibility mechanism, the fund still might not live up to the expectations of EPA, the state, or the UST owner/operator.

EPA review of state funds as financial responsibility mechanisms includes four main elements:

- Funding source
- Amount of fund
- Coverage provided
- Eligibility for use of the fund

10.2.1 Funding Source

Funds provided by the state for clean-up and third-party damages must be "reasonably certain and available." Although it is not entirely clear how EPA determines whether funds are "reasonably certain and available," some guidelines are given. According to EPA, a state fund that depends on annual appropriations out of the state's general revenues from its legislatures would *not* adequately assure that funds would be "certain and available." In addition,

---

* This section is based on: U.S. Environmental Protection Agency, *Reviewing State Funds for Financial Responsibility,* enclosure in letter from Helen Lundsford, U.S. Environmental Protection Agency, Region IV, to Paul Thompson.

the federal LUST Trust Fund cannot be used for this purpose either. Acceptable sources of funds include petroleum taxes, licensing or UST fees, bond issues, and risk-based premiums. The fund does not have to be used solely for UST releases.

10.2.2 Amount of Fund

EPA does not establish a particular formula for determining how much money should be maintained in a state fund. Instead, EPA suggests that funds should be thought of as:

> a "bank account", with money being "deposited" and money being "spent" as it is needed. The goal here is to reasonably assure that the projected flow of revenues into the fund is sufficient to keep pace with the anticipated rate of expenditures from the fund.

The guidance does not specify that the fund generate enough revenue to achieve a particular degree of progress in cleaning up existing leaks.

Because state funds provide so many different kinds of coverage, the amount of funds needed can differ. EPA does suggests the use of four mechanisms in a state fund to guarantee that adequate funds are maintained. These four mechanisms include:

- A triggering provision to allow the funding source to be activated once the level of the fund reaches a predetermined *minimum limit*
- A triggering provision to allow the funding source to be deactivated when the level of the fund reaches a predetermined *upper limit*
- A triggering provision to allow the additional collection of funds when a state expects that a large release UST release will be a significant drain on the state fund
- For states using fees to support their funds, a provision to allow for the modification of the fee structure

10.2.3 Coverage Provided

State funds generally can provide for either full or partial coverage. *Full coverage* funds allow *all* UST owners/operators in the state to comply *fully* with the federal financial responsibility requirements *without any reliance on other financial responsibility mechanisms*, such as insurance or a state-required mechanism (for example, a financial test to demonstrate financial responsibility for a deductible). Full coverage funds may even go *beyond* the federal financial responsibility requirements by not limiting the coverage provided by the fund to the amounts set in the financial responsibility regulations. Thus, the state fund could cover the full amount of corrective action and third-party compensation required by an UST release, even if the total cleanup costs were above the amount of financial responsibility required by EPA. This description of full coverage funds can be misleading, however, because such funds can still require UST owners/operators to pay deductible amounts (even deductibles of $100,000) as long as the fund provides for *first dollar coverage*. First dollar coverage means that the state fund can be used in cases where the UST owner/operator is unable or unwilling to pay the required deductible amount. The state would then have the discretion of pursuing a cost recovery action against the UST owner-operator. Thus, full coverage funds do not necessarily (and rarely do) remove all responsibility of corrective action and third party compensation from the UST owner/operator.

There are two types of *partial coverage* funds that EPA will approve. In addition, EPA can approve state funds that combine features from both. The first type only provides coverage for a *portion of the dollar amounts or types of coverage* (corrective action or third-party liability) required by EPA. Under these types of state funds, UST owners/operators must obtain an acceptable financial responsibility mechanism (such as insurance or a state-required mechanism) for the dollar amounts or types of coverage not provided by the state fund. This additional financial responsibility mechanism is called "wrap-around" coverage.

Thus, in contrast to the full coverage fund, the partial coverage fund does not require the state to provide *first dollar coverage* and the UST owners/operators must obtain another mechanism to demonstrate financial responsibility for the deductible amount not covered by the state. (Although under the full coverage fund the state does not require that deductible amounts be guaranteed by a financial responsibility mechanism, the state may still pursue a court action to recover from the UST owner/operator the amounts spent by the state within the deductible limits.)

Because under this type of partial coverage fund the state assumes *some* financial responsibility for an UST release, EPA allows states to establish their own financial test of self-insurance for deductible amounts not covered by the fund. EPA guidance suggests that the financial test require that an UST owner's/operator's minimum net worth be a specific multiple of the deductible amount.

The second type of partial coverage fund provides coverage for only *some owners or operators* in the state. These types of funds require those UST owners/operators not covered by the state to demonstrate the full amount of required financial responsibility using some mechanism other than the state fund.

States that provide either full or partial coverage must provide "reasonable assurance" that it will pay the covered amounts of eligible UST owners/operators in the case of an UST release. This assurance can be provided in one of three ways. First, the state may undertake corrective action at the site and pay for cleanup and third party compensation directly, usually only in instances where the UST owner/operator is unable or unwilling to pay these costs.

Often, however, state funds are designed so that the responsible party (RP) — usually the UST owner/operator — will undertake the cleanup, either voluntarily or pursuant to a state administrative or judicial order. In this type of fund, the state then would pay the RP or his contractor for covered expenses as corrective action and third-party compensation activities progress.

The third approach pays out costs only after the UST owner/operator has paid for the cleanup and compensated third-parties, and applied for reimbursement along with proof of the expenses incurred. *EPA will not approve the use of these types of funds unless they also will cover costs that the owner or operator is unable or unwilling to cover before being reimbursed.*

10.2.4 Eligibility for Use of the Fund

State funds approved by EPA can provide either *uniform or unlimited* eligibility. Unlimited eligibility means that the state fund covers *all* UST owners/operators in the state. Funds with limited eligibility establish "entrance requirements" that restrict the use of the fund only to those UST owners/ operators that meet certain conditions, such as the performance of a tank-tightness test. States that limit eligibility should inform any UST owners/ operators in advance that they are not eligible for the fund.

State funds approved by EPA may limit eligibility to UST owners/operators that are in "substantial compliance" with federal and state UST technical requirements *at the time of the release*. Under these types of funds the state, much like an insurance company, determines after an UST release occurs whether the UST owner/operator was in compliance with any applicable technical requirements, and thus eligible for coverage. Because some UST owners/operators may not be aware before a release occurs that they were not in "substantial compliance" with federal and state regulations, EPA suggests that eligibility criteria be specific enough so that owners/operators can determine ahead of time what they must do to qualify for coverage. EPA also "strongly urges" — but does not require — states to allow the use of the state fund (followed by cost recovery if necessary) even if questions about the eligibility of the UST owner/operator arise.

Finally, EPA can approve state funds that are set to expire at a designated time in the future (the so-called "sunset provision"), presumably after the UST insurance market has expanded or UST upgrading has taken place. Approval of funds with sunset provisions is limited to the time for which the fund is currently authorized, or until it ceases to provide the required levels of coverage. The state must notify all covered UST owners/operators that the fund's coverage is terminating at least 60 days before the fund becomes inactive.

10.3 POTENTIAL PROBLEMS WITH STATE FINANCIAL RESPONSIBILITY FUNDS

Chapter 7 indicated that the most effective state financial responsibility funds will tend to limit the public's liability for existing UST releases and provide adequate incentives for tank upgrading and sound UST management practices. An overly broad assumption of public liability and a lack of incentives for the UST owners/operators could lead to very high levels of public funding and political opposition, both of which could threaten the viability of the state fund as a financial responsibility mechanism. Even state funds that conform to these criteria still suffer from inherent conflicts stemming from the dual role the state must play in protecting public funds from misuse and safeguarding human health and the environment.* The development of an UST insurance market able to provide affordable coverage for large segments of the regulated community also could be inhibited by the use of state financial assurance funds.

These problems can be understood by examining three specific characteristics of state funds, which include: (1) levels of state funding and fund financing; (2) eligibility requirements; and (3) claims management. In addition, some states are attempting to reduce the liability of the state fund by offering low-interest loans for UST upgrading. Unfortunately, these loan programs could also result in excessive public liability for costs associated with operating USTs. This section examines these three fund-characteristics, as well as the use of state loan programs for UST upgrading, in greater detail.

10.3.1 Levels of State Funding and Fund Financing

State funds usually provide between $3 million and $30 million annually to cover UST releases. These amounts may be far too low, especially for state funds that (1) provide full financial responsibility coverage both for corrective action and third-party compensation and (2) cover all sites within the state regardless of whether they have been certified as uncontaminated or can qualify for and afford other financial responsibility mechanisms.[1,2]

For example, assuming that an average UST leak incurs $100,000 in corrective action and third-party liability costs would mean that a state fund with $10 million annually could only perform approximately 100 cleanups a year. Many states, however, have thousands of leaking USTs which would take decades to address with a fund of this size. Clearly, a private UST insurance company could never provide coverage for such a large and risky group of clients as *all existing USTs in an entire state* with such a limited pool of capital. Based on the figures above, it will take over $30 billion just to address the 200,000 *existing* UST leaks that EPA estimated in 1988.[3]

Perhaps the best example of this problem is the "amnesty" program established as part of Florida's trust fund to pay for the cleanup of contaminated sites. UST owners/operators with leaking tanks could apply for and receive up to $1 million per site for corrective action. Funding levels for the $50-million fund were based on estimates that there would be approximately 2000 applications for site restoration.[1,4] By the time the amnesty period ended in 1988,

---

* See Chapter 7, Potential Conflicts in Administering State Funds.

the state received almost 9,400 applications out of a total of 29,000 facilities.[1,4] The state Department of Environmental Regulation estimated in 1989 that it would be able to conduct approximately 100 cleanups each year, with costs for existing cleanups averaging $188,000.[1] Thus, the state exposed itself to almost $16 billion in liability and could be in the business of cleaning up USTs well into the foreseeable future.

In addition to the problem of under-capitalization, states that only provide financial responsibility for corrective action put UST owners/operators in the position of obtaining coverage for third-party injuries. Even if this type of wrap-around coverage was offered (it is not widely available), many firms that cannot obtain or afford insurance for both corrective action and third-party injuries also will be unable to do so for only third-party injuries. (Wrap-around coverage is discussed further in the following section.)

Finally, state funds may not have mechanisms to ensure that incoming revenue will equal outgoing expenditures based on expected annual leak discoveries and cleanup activities. Under these circumstances, insurers may be reluctant to provide coverage in conjunction with a state fund because of fears that they will be held liable for costs intended for the state fund but for which no state dollars are available.[1,2,5] After all, the contractors, lawyers, and other professionals involved in the site cleanup will expect to be paid for the work performed regardless of who (i.e., the state or the insurance company) makes the payments.

10.3.2 Eligibility Requirements

A closely related problem is that funds may provide overly-broad coverage. As mentioned above, state financial responsibility funds that cover all tanks and all associated cleanup and third party compensation costs run the risk of inheriting broad liability for thousands of previously contaminated sites or USTs that should be upgraded. Overly generous eligibility requirements not only provide coverage to firms that do not need state assistance in meeting financial responsibility regulations or those that represent unacceptable risks, but also remove financial incentives for upgrading USTs or engaging in good tank management procedures.[1,2,5,6]

Unlike insurance, the fees required from UST owners/operators for coverage by a state fund often are not "risk based" or set to reflect the actual cost associated with the coverage provided. Instead they may be kept artificially low through state subsidies and require only a single fee (such as an UST registration fee) regardless of the different risks presented by a separate firms.[1,2,5]

The deductibles charged by state financial responsibility funds also pose potential problems. Many states set the deductibles at relatively low levels — often $10,000 or less. Again, the incentive to upgrade tanks and engage in good UST management are reduced significantly if the deductible is too low, especially if the state fund does not levy risk-based premiums. On the other hand, some states set deductibles at levels of up to $200,000. (And as mentioned above, other funds only cover costs associated with corrective action — not third party liability.[1,2,5]) Unless the state fund provides "first dollar coverage", many smaller firms would still have to obtain private wrap-around coverage (i.e., insurance for the amount of financial responsibility required by EPA that is not provided through the state fund) for the amount of the deductible.[1,2,5]

10.3.2.1 The Use of a State-Required Financial Test to Demonstrate Financial Responsibility Deductibles Under a State Fund

As discussed in Chapter 8, EPA's financial responsibility regulations authorize states whose funds do not provide first dollar coverage to allow UST owners/operators to use a "financial test" to demonstrate financial responsibility for the amount of the deductible. (EPA considers such a financial test to be a "State Required Mechanism," pursuant to 40 CFR Section 280.100). Although these tests usually are not as stringent as the federal financial test of self-insurance (which is used to demonstrate the *entire* amount of financial responsibility), many smaller UST owners/operators may not have sufficient net worth to pass a state's financial test for a high deductible.

In addition, although wrap-around coverage is available in some areas, it still is difficult to obtain and can be almost as costly as full coverage. Insurance companies will remain reluctant to offer coverage to high-risk firms with older USTs and poor tank management procedures. And since most claims are in the $100,000 range, insurance premiums may not be substantially lower than without the fund if UST owners/operators must obtain insurance to cover high deductibles required under the state fund.[1,2,5-7] Thus, availability and affordability are still issues for private coverage that is offered in conjunction with state financial responsibility funds.

At any rate, UST owners/operators should not be overly optimistic about demonstrating financial responsibility for the deductible amounts through a lenient state "financial test" that is designed to make it easier for firms to comply with EPA's coverage requirements. While the state fund and the state financial test may allow UST owners/operators to more easily *demonstrate* the full amount of financial responsibility required by EPA, in the event of a leak the owner/operator must still *pay* the deductible in full.

Therefore, the UST owner/operator must be sure that adequate funds are available to pay the deductible even if the state required mechanism might allow him to "slip by" without the necessary financial capability. And because of the leak detection and reporting requirements that are part of EPA's UST technical regulations, UST owners/operators should assume that existing leaks will be detected more frequently than before the regulations.

Finally, UST owners/operators covered by a state fund that offers first dollar coverage should not assume that all liability associated with an UST release will be assumed by the state merely because they are not required to obtain a financial responsibility mechanism to cover a deductible. Many state funds with first dollar coverage still require that the UST owner/operator pay a deductible and authorize the state to recover any funds expended within the deductible limits in cases where an owner/operator refuses to cooperate.

10.3.3 Claims Management

State funds often are administered by officials within the state's environmental agency, which may be understaffed and lacking in any real expertise with managing insurance claims. In addition, state regulators may not have the flexibility to change underwriting criteria and premium structures to minimize the chances of a leak at a particular facility. Thus, given the number of LUSTs, state administration of claims could significantly increase costs and reduce environmental protection.[1,2,5]

Finally, state financial assurance funds must address the question of whether the state or the insurance company should oversee the cleanup of UST leaks in situations where the company provides wrap-around coverage. Insurance companies may be unwilling to provide this partial coverage to firms also covered under a state fund if the insurer is required to give up its traditional right to oversee the cleanup process. In addition, companies providing wrap-around UST insurance could be reluctant to manage an entire corrective action or third-party claim and then wait months, possibly years, for reimbursement by the state fund.[1,2,5]

10.3.4 State Programs for Tank Upgrading

One important problem faced by state officials considering the use of a state fund is how to provide coverage for sites that are already contaminated or sites with very high-risk tanks. Such coverage certainly does not make sense for the private insurer and state assumption of liability for these USTs could mean enormous expenditures. Some 13 states have established low interest loan programs for UST upgrading to be used before financial responsibility requirements are enforced.[8] After the tanks are upgraded, the UST owner/operator could participate in the state fund (or obtain private insurance) because the USTs no longer represent an unacceptably high-risk. And given that use of the fund for financial responsibility could be conditioned on good management practices, the tanks would remain in good condition as long as they were covered by the state fund.

This approach has its merits, but such loans undoubtedly would accelerate the discovery of leaks, making the need for financial responsibility all the greater. An UST upgrade loan program without some sort of financial-responsibility requirements could open the door to state assumption of massive liability for tank closure, corrective action, and third-party compensation *before UST upgrading could even begin*. State assumption of this responsibility, just as with an overly-broad state financial-responsibility fund, could require massive infusions of public resources or lead to potential conflicts of interest. To date, few states have passed comprehensive loan programs for UST upgrading.

Requiring UST owner/operators to assume this liability for cleanup costs that are not directly related to UST upgrading would be a disincentive to seek state loans for upgrading because the owner/operator would face another "catch 22", similar to the one encountered when seeking a private bank loan to qualify for insurance (discussed in Chapter 7 and 9). In this case, in order to comply with financial-responsibility requirements UST owners/operators must obtain a loan to upgrade their tanks, but upgrading USTs can lead to the discovery of a leak, in which case there is no financial-responsibility mechanism to provide the necessary cleanup funds. Thus, loan programs for UST upgrading may be feasible only when the loan recipient can certify that the site is clean or that state funds expended for site cleanup can be repaid.

10.3.5 Summary of Questions about State Financial Assurance Funds

It is apparent from the preceding discussion that although all state financial responsibility funds share some things in common, including certain shortcomings associated with any public assumption of private liability for UST leaks, such funds can differ in many ways. These differences can make one fund a viable solution to EPA's UST financial-responsibility regulations while leaving another questionable, at best. An UST owner/operator must answer many questions about a particular state fund — or fund proposal — in order to understand whose interests the fund serves. UST owners/operators also need to determine the likelihood that the fund will be successful either in providing an effective, longterm compliance option for the UST financial-responsibility regulations or fostering the development of a larger and more affordable UST insurance market.

Unfortunately, merely answering these questions "yes" or "no" will not indicate how useful a particular state fund will be to UST owners/operators. This is because UST owners/operators (who compete with each other) do not have the same financial conditions or UST upgrading requirements. In addition, some UST owners/operators only need short-term coverage under a state fund (e.g., until insurance can be obtained or USTs are closed) while others would prefer to depend on the state to meet its long-range UST financial-responsibility requirements. Clearly, UST owners/operators who can afford to upgrade their tanks and obtain UST insurance may not want an overly generous state financial-responsibility fund or loan program to provide subsidized coverage to its competition. Finally, UST owners/operators must consider how other interested parties, namely the UST insurance industry and the public, would view the answers to these questions, since the success of a state financial responsibility fund (and of EPA's financial responsibility program) is influenced by their interests.

To understand the advantages and drawbacks of various types of state financial assurance funds, this section summarizes some of the questions that UST owners/operators should ask.

10.3.5.1 General Fund Characteristics

1. Will the state fund provide the amount of money needed to clean up all of the existing UST leaks covered by the fund during EPA's UST upgrading period (by 1998)? Obviously, those UST owners/operators covered by the fund — especially those who have no other viable options for obtaining a financial-responsibility mechanism until their USTs are upgraded should be concerned that the state fund provide sufficient capital for all covered leaks. Depending on the scope of coverage provided and the stringency of eligibility require-ments, however, the UST insurance industry and the public may oppose state funds that result in high levels of public expenditures — especially indis-criminate expenditures.
2. Does the state fund have a mechanism to increase revenue levels in case the state underestimates the amount of money needed to address existing leaks? Again, to the extent that an UST owner/operator depends on the coverage provided by the fund, it is important that the state have a mechanism to increase funding levels when needed, although the insurance industry and general public both have interests in reducing the state's liability for leaking USTs.

10.3.5.2 Third-Party Coverage

3. Does the state fund provide for both corrective action and third-party compen-sation? Funds that do not provide for third-party coverage require UST owners/operators to obtain a separate financial responsibility mechanism for the *full amount* of financial responsibility (up to $2 million of annual aggre-gate coverage).
4. If third-party claims are covered, does the state issue payments from the fund pursuant to judicial decisions arising out of third-party civil suits? State funds that pay third-party claims based on awards that are determined through civil suits may be attractive to the UST owner/operator but would not satisfy the public's need to have its funds controlled by the state.

10.3.5.3 Scope of Coverage and Eligibility Requirements

5. Does the state fund apply to all UST owners/operators in the state or only to certain UST classes or categories? Funds that cover all tanks regardless of the financial characteristics of the UST owner/operator are not in the best interest of small firms with limited resources. State financial responsibility funds that cover firms that could otherwise obtain a financial responsibility mechanism (such as insurance) result in unnecessary public expenditures and deter the development of UST insurance markets, both of which threaten

the longterm viability of the fund for those firms with a legitimate need for state coverage.

6. Does the state fund have eligibility requirements to cover only uncontaminated sites or sites with USTs that are not leaking? State funds that cover existing contaminated sites obviously will appeal to those firms with LUSTs but not to firms with clean sites. The increased public expenditures and disincentives for proper UST management could threaten the fund by leading to public opposition and deterring UST insurance markets.

7. Does the state fund have eligibility requirements to cover only USTs that meet certain upgrading requirements? State funds without eligibility requirements for tank upgrading will serve those firms that cannot afford upgrading but are not in the best interest of other firms. Equal coverage for all USTs, regardless of UST characteristics will result in increased public expenditures and reduce the availability of UST insurance by reducing incentives for tank upgrading and altering the characteristics of the UST universe.

10.3.5.4 State Loan Programs for UST Upgrading

8. If the state fund limits coverage only to upgraded USTs, is it accompanied by a loan upgrade program? State loan programs for upgrading tanks may provide many UST owners/operators with the assistance they need for longterm compliance with the financial responsibility regulations. The public, as well as UST owners/operators genuinely in need of state assistance, may react negatively to programs that offer loans to firms already possessing adequate financial resources or credit to upgrade tanks.

9. If the state has a loan program for upgrading USTs, does it apply to site remediation as well as to tank upgrading? Many firms that cannot afford to cleanup existing UST leaks would prefer that a state loan program for tank upgrading also provide funds for corrective action and third-party compensation. The public would tend to oppose state loans used for these purposes if they resulted in excessive public expenditures. Such a generous loan program also may not be in the best interest of UST owners/operators who need a loan for tank upgrading but do not have LUSTs, as well as owners/operators who have already upgraded their tanks using their own resources.

10.3.5.5 First Dollar Coverage and Deductibles

10. Does the state financial responsibility fund provide for "first dollar coverage?" First dollar coverage will appeal to UST owners/operators who cannot demonstrate financial responsibility for even the deductible portion required under the state fund. While the state may seek reimbursement from UST owners/operators for this deductible, first dollar coverage will lead to increased public expenditures because of cases where UST owners/operators refuse — or are unable — to reimburse the state for the deductible amount but who still qualify for coverage under the state fund. In addition, first dollar coverage might deter UST insurance since it could reduce incentives for good UST management practices (including timely UST upgrading) and lead to

coverage for firms without the financial resources to pay deductibles. Both of these impacts could alter both the financial and UST management characteristics of UST-owning firms, which are used by the insurance industry to gauge the risks of providing coverage in a given area.

11. Does the state financial responsibility fund require UST owners/operators to pay large deductibles? Generally, UST owners/operators would prefer state financial responsibility funds that require only small deductibles — especially firms without adequate financial resources to pay larger deductibles. Small deductibles are not as beneficial to larger firms because they require larger public expenditures and could deter UST insurance availability by decreasing incentives for good UST management practices (including tank upgrading).

10.3.5.6 Wrap-Around Coverage

12. If the state financial responsibility fund does not provide first dollar coverage, what financial responsibility mechanisms may be used for wrap-around coverage? Financial responsibility mechanisms for wrap-around coverage that do not require significant net worth have an obvious appeal to UST-owning firms with limited financial resources. Larger firms that could demonstrate financial responsibility for wrap-around coverage under a more rigorous financial test (or even obtain insurance for the amounts not covered by the state fund) do not necessarily benefit from more lenient state-required mechanisms. This is because lenient state-required mechanisms will provide coverage to firms that may not be able to cover deductibles — which will increase the need for public expenditures; and will alter the financial and UST characteristics of firms in the regulated community — which will decrease UST insurance availability.

13. Does an insurer providing wrap-around coverage have the right to oversee that portion of the cleanup that he/she covers? While the state may choose to maintain control over the entire cleanup process, it is in the best interest of both UST owners/operators and the UST insurer if the insurer providing wrap-around coverage oversees that portion of the cleanup he/she covers. The UST insurance company can most efficiently manage its own funds and can pass any savings on to the UST owner/operator through reduced premiums. The public, on the other hand, has an interest in guaranteeing that the insurer abide by state and federal corrective action requirements and cleanup standards. The most effective compromise is to allow the insurer to manage the cleanup process subject to oversight and approval by state officials.

10.4 CURRENT STATUS OF STATE FINANCIAL ASSURANCE FUNDS

States have been developing UST financial assurance funds since EPA issued the financial responsibility regulations in October, 1988. As of April

1991, 46 states had established funds although EPA has approved only 27 state funds as of January 1992.[8,9]

While EPA has not approved as many state funds as anticipated initially, in recent months the pace of approval has increased. EPA approval of state financial-responsibility funds remains slow, however, with only 12 funds being approved between May 1990 and April 1991* and only six more funds approved by January 1992. Since the financial responsibility compliance deadline already has passed for UST owners/operators with 13 to 99 tanks, however, the time for approving state financial responsibility funds is running short. (As discussed in Chapter 8, once a state fund is submitted for EPA approval, UST owners/operators are deemed to be in compliance with the amounts and types of costs covered by the fund pending a final determination by EPA.)

Table 6 (found at the end of this chapter), prepared by the Petroleum Marketers Association of America, provides some indication of the most common types of state funds. The majority of state financial responsibility funds provide $1 million of coverage, thus providing the financial responsibility needed by UST owners/operators with less than 100 tanks. However, up to $2 million in annual aggregate coverage is provided by 12 funds. Three states, Arizona, North Dakota, and Wisconsin, provide coverage levels of less than $1 million.

Most funds cover both corrective action and third-party compensation although 11 of the funds only cover corrective action or do not provide coverage for all third-party damages (e.g., the Missouri fund only covers third-party property damages, not injuries). All but five of the state financial responsibility funds require UST owners/operators to participate in the fund. Most of these funds, therefore, provide coverage to virtually all UST owners/operators regardless of their ability to afford alternative financial responsibility mechanisms.

Only five of the state funds provide incentive programs to encourage tank upgrading, while virtually none of the funds charge risk-based premiums to generate the appropriate levels of capital and provide incentives for sound UST management. Many funds, however, generate capital — at least in part — through one-time or annual tank fees of $5 to $500, with most fees falling in the $50 to $150 range. Deductible limits vary widely, but many funds set deductibles of $5,000 to $25,000.

10.4.1. Description of Selected State Funds

Because it is difficult to generalize about specific state financial assurance funds, this section briefly discusses four funds — each of which have been approved by EPA. While these funds are not representative of all state financial responsibility funds, they do indicate some of the types of financial responsi-

---

* This figure is based on a comparison between the 21 states approved as of April 15, 1991,[8] and the 9 states approved as of May 2, 1990, according to 53 FR 18567.

bility coverage that can be provided by states under EPA regulations and how much this coverage can differ. UST owners/operators interested in specific state funds should contact the appropriate state UST offices. A list of these offices is included in Appendix B.

10.4.1.1 The Michigan Underground Storage Tank Financial Assurance Fund (MUSTFA)*

Michigan's financial-responsibility fund (MUSTFA), administered by the state Department of Management and Budget, was authorized to begin disbursing funds to covered owners/operators on February 15, 1990. MUSTFA is capitalized by a fee of .875¢/gal levied on refined petroleum products that are sold for resale or consumption in the state. The fees are collected generally at the same time that sales taxes are collected. The fund covers corrective action and third-party liability costs up to $1 million.

MUSTFA can be used for any corrective action and third-party liability associated with leaks discovered and reported after July 18, 1989. To use the fund, an UST owner/operator must be in compliance with applicable state and federal UST technical requirements, which include reporting and recording keeping, inventory control or leak detection, reporting of suspected releases, and corrective action and closure requirements. A deductible of $10,000 must be met before the fund can be used for which financial responsibility must be demonstrated.

Within 3 years after the fund becomes operational, UST owners/operators must comply with the federal** and state standards for new USTs in order to receive coverage for third-party damages. Once an owner/operator has received money from the fund, he is no longer eligible to use the fund unless he upgrades or replaces the LUSTs so as to comply with the technical standards for new USTs. Within 2 years after the effective date of the fund (July 18, 1989), up to 10% of the fund's annual revenues can be used to provide interest subsidies to lenders providing loans to UST owners/operators for tank replacement.

UST owners/operators who experience an UST leak must prepare a corrective action plan that complies with state regulations and submit a bid to the fund's administrator for implementing the plan. If the administrator determines that the bid is reasonable and that the UST owner/operator is eligible, the owner/operator can submit work invoices to the administrator for approval. *If sufficient money exists in the fund,* the administrator will make payments to a contractor or to the owner/operator for work approved by the state Department of Natural Resources as consistent with the corrective action plan. The state

---

* Unless otherwise noted, the material in this section is derived from: Michigan UST Financial Assurance Act (Act 518 — P.A. of 1988) as amended by Act 152 of 1989; UST News, Michigan State Police, Vol. 1, No. 3; Vol. 2, No. 1; Vol. 2, No. 2; and Vol. 2, No. 3.
** See, 40 CFR Sections 280.20.

Attorney General must approve requests by owners/operators that the fund be used to pay third-party claims (including court judgments and "out-of-court" settlements).

The state is not required to pay for approved work invoices and approved third-party claims if there is insufficient money in the fund to make a payment. Pending a review of UST insurance availability, MUSTFA is scheduled to expire 5.5 years from its initiation in 1989. Although not based on an actuarial study, the necessary size of the fund (currently estimated at $60 million) is based on total payouts from previous years. The fund had covered the costs (averaging $65,000) associated with leaking USTs for approximately 150 firms as of September, 1991.*

10.4.1.2 The Louisiana Underground Motor Fuel Storage Tank Trust**

Louisiana's state financial assurance fund is operated by the state Department of Environmental Quality and originally was capitalized through tank fees not to exceed $150 per tank. The fund can never drop below $6 million or exceed $20 million. Recent state legislation amended the source of capital for the fund by providing for a "transfer fee" of .15¢/gal for fuel deliveries, which is collected by the state from bulk dealers, who in turn collect the fees from retailers.†

Any UST owner/operator that has registered their tanks with the Department, paid the appropriate tank fees, and met the financial-responsibility requirements for the amounts not covered by the fund, are eligible to use the fund. (Owners/operators of unregistered tanks must first demonstrate that the tanks are not leaking and have never leaked.) The Department reserves the right to seek reimbursement for any funds expended by the state for a LUST, if the Department determines that the tank was not in *substantial compliance* with all applicable state and federal UST laws and regulations when the leak occurred.

The fund must be used whenever the department determines that a LUST (owned or operated by an eligible fund participant) "may pose a threat to the environment or the public health, safety, and welfare." It is not clear whether this determination corresponds to the federal requirement that UST owner/operators begin corrective action once a release is detected.

UST owners/operators must pay deductibles of $15,000 for corrective ac-

---

*   Chris Little, Michigan Department of Management and Budget, personal conversation with Paul Thompson, February 27, 1991.

** Unless otherwise noted, the material in this section was derived from enclosures in correspondence from Patsy Deaville, Louisiana Department of Environmental Quality, UST Division, to W. David Conn, December 5, 1989.

†   George Gullet, Louisiana Department of Environmental Quality, Underground Storage Tank Division, personal conversation with Paul Thompson, February 27, 1991.

tion and $15,000 for third-party damages before the state fund can be used. Owners/operators must demonstrate financial responsibility for these amounts. The legislation establishing the fund does not establish a time period when leaks must have occurred or been reported in order to receive coverage under the fund for corrective, although payments by the fund to damaged third parties can be made only if the leak occurred while the UST owners/operator was an eligible fund participant.

Third-party claims must be made against the UST owners/operator as well as the state, in its capacity as administrator of the fund. The state attorney general is required to act in behalf of the state as a representative of the fund. The costs to the state of defending the fund against third-party claims must be recovered from the fund.

Like most other state financial responsibility funds, the Louisiana fund limits its liabilities for UST cleanup and third-party compensation. The state has no "liability or responsibility to make any payments for cleanup costs or third party claims if the (fund) ... is insufficient to do so." Of the 24,850 USTs registered in the state, the fund had paid approximately 17 claims as of September, 1991.*

10.4.1.3 The Georgia Underground Storage Tank Trust Fund**

Georgia's state financial responsibility fund is operated by the state Department of Natural Resources and is financed by an "Environmental Assurance Fee" (EAF) of .1¢/gal of petroleum products paid by UST owners/operators when they purchase the products from distributors. Participation in the fund is not mandatory, but those UST owners/operators that sought coverage by the state as of July 1, 1988 must have paid the EAF and reported any known or suspected leaks from an existing UST.

Owners/operators with existing USTs who wished to participate in the state fund 90 or more days after July 1, 1988 must have submitted proof that the tanks passed a "tank tightness test" along with the EAFs which would have been collected had they elected to participate in the fund initially. Coverage by the state fund for new USTs requires that the tanks be installed in compliance with EPA regulations for new USTs. Regardless of when the UST owner/ operator began participating, the state fund cannot be used when the state determines that a release existed before July 1, 1988.

To maintain eligibility, UST owners/operators must report any changes in

---

* George Gullet, Louisiana Department of Environmental Quality, Underground Storage Tank Division, personal conversation with Paul Thompson, February 27, 1991.

** Unless otherwise noted, the material in this section was derived from enclosures in correspondence from Marlin R. Gottschalk, Ph.D., Georgia Department of Natural Resources, Underground Storage Tank Unit, to W. David Conn, January 9, 1990.

UST status (including changes in ownership or closure), pay EAFs, and maintain any required records (including the results of any tank tightness tests). The Department of Natural Resources must issue a notice of ineligibility to UST owners/operators that do not correct violations of these requirements within 30 days. Owners/operators declared ineligible for use of the fund then have an additional 30 days to obtain an alternate form of financial responsibility.

The state financial responsibility fund provides $1 million per-occurrence coverage for all eligible UST owners/operators. In addition, the fund provides $1 million annual aggregate coverage for owners/operators with less than 101 USTs and $2 million for UST owners/operators 101 USTs or more. A deductible totaling $10,000 both for corrective action and third-party liability must be paid before the fund can be used. UST owners/operators must demonstrate financial responsibility for this amount. The fund also can be used by the state for emergency cleanups where public health and safety are threatened and in situations where the owner/operator of a LUST cannot be identified or is unwilling or unable to perform corrective action.

Eligible UST owners/operators with corrective action plans approved by the state must submit an application for reimbursement within 30 days of completing the corrective action. No costs may be reimbursed by the state before the corrective action has been completed unless provisions for interim payments have been made and the corrective action is being completed in accordance with the approved cleanup plan. The state fund can only be used for third-party claims if the UST owner/operator gives the state 60 days notice that a claim has been made against him. Payments by the fund to injured third-parties will be made only when the Department of Natural Resources: (1) approves a certification from the UST owner/operator and the injured third party (along with their attorneys) that a third-party claim should be paid and; (2) receives a valid final court order establishing a judgment against an eligible UST owner/operator for third-party damages.

Although the current fund is limited to tanks that have not leaked, the state is considering — under pressure from oil jobbers, convenience store owners, and independent retailers — raising the Environmental Assurance Fee and expanding the fund's coverage to *all* USTs in the state, regardless of whether they are leaking. This would put a significant strain on state resources, since there are 51,400 USTs registered in Georgia and over 1,300 reported UST releases, with many other leaks still not discovered.*

As of September 1991, the state fund had approved three claims, with one claim of $67,141 being paid. As the fund's primary purpose is to allow UST

---

* Marlin Gottschalk, Georgia Department of Natural Resources, Underground Storage Tank Unit, personal conversation with Paul Thompson, March 3, 1991.

** Marlin Gottschalk, Georgia Department of Natural Resources, Underground Storage Tank Unit, personal conversation with Paul Thompson, March 3, 1991.

owners and operators to comply with the financial responsibility regulations, no actuarial study was conducted to determine the necessary size or administrative aspects of the fund to provide the liability coverage needed by the UST-owning community.**

10.4.1.4 The Idaho Petroleum Clean Water Trust Fund†

The state fund used in Idaho differs from most others because it is an *insurance* — rather than *assurance* — fund and is operated by the State Insurance Fund, the agency that runs the state's workers' compensation fund. According to the state legislation establishing the program, the fund is "a liability insurance trust fund that will make contracts of liability insurance available to owners and operators of petroleum storage tanks ... through fair and equitable insurance contracts issued by a state-licensed nonprofit organization."

The fund is administered much like a private insurance company and is charged with providing underwriting, claims handling, and actuarial services. The fund must be registered with the state Department of Insurance and comply with most applicable state insurance laws. The Department of Insurance cannot register the fund to offer insurance within the state unless it determines that it is "actuarially sound" (i.e., its assets, income, and other financial resources are adequate under reasonable estimates for payment of all claims and other expenses).

Current estimates are that $10 to $20 million will be needed to operate the program. Funding for the program is generated by a "transfer" fee of 1¢/gal on fuel products and fees of $25 per tank that must be paid by UST owners/operators for the state's UST insurance coverage.

The key to ensuring that the fund's revenues match its expenditures is excluding any USTs from coverage that are already leaking. Before USTs can be insured by the state, the owner/operator must certify that they are in compliance with federal, state, and local UST regulations and have passed tank tightness tests. In addition, the UST owner/operator must certify that any existing contamination has been — or is being — cleaned up.

It is likely that only about 6000 USTs out of a state total of 9000 will qualify for the state insurance program. Although this approach does not help UST owners/operators with existing contamination, it does provide relief to small firms with sound USTs that cannot afford private insurance as well as firms with older tanks that may not qualify for private insurance.

It is still to early to gauge the success of Idaho's Petroleum Clean Water Trust Fund (applications for coverage are currently being sent to UST owners/

---

† Unless otherwise noted, the material in this section was derived from enclosures in correspondence from Richard Ostrum, Idaho Petroleum Storage Tank Fund, to Paul Thompson, January 4, 1991.

operators), but it could provide many firms the coverage needed not only to comply with the financial-responsibility regulations, but also to ensure longterm protection against LUST liability. For owners/operators that qualify for coverage, the fund generally is implemented like a private UST insurance program. The fund provides for "first-dollar" coverage, but a $10,000 deductible is required. Coverage levels and provisions are consistent with the levels established in EPA's financial responsibility regulations. Although the state program does suffer some drawbacks, notably the absence of risk-based premiums, it goes a long way to eliminating many of the disadvantages of other state funds, even if it does so by imitating private insurance.

10.5 CONCLUSION

The use of state financial assurance funds is the only means by which most UST owners/operators will be able to comply with the federal UST financial responsibility regulations. In many cases, these funds can provide the coverage needed to comply with the financial responsibility regulations. To be considered successful, however, state funds must either provide a longterm compliance option or foster (or, at the very least, not inhibit) the development of affordable UST insurance. In addition, the funds must be viewed not merely as techniques for complying with regulations, but as viable tools for allowing UST-owning firms to remain in business after an UST leak. A stable financial responsibility mechanism is all the more important now that leak detection is an important feature of EPA's overall UST program.

Unfortunately, the vague nature of the guidance EPA uses to approve state funds can lead to financial responsibility coverage that is less reliable than coverage provided by private UST insurance. And some state funds may still require UST owners/operators to shoulder significant responsibility for tank leaks by requiring the payment of large deductibles or mandating that additional "wrap-around" coverage be provided. In addition, inadequate capitalization levels, overly generous eligibility criteria, and poor claims management can threaten the viability of the fund by straining public resources and deterring the private UST insurance market. UST owners/operators should carefully consider whose interests are served the most — and whose are served the least — by the particular characteristics of a state fund.

Many states already have established funds and EPA is stepping up its efforts to approve these funds before any more compliance deadlines pass. Generally, most funds have low deductibles and provide the coverage needed for firms with less than 100 tanks to comply with the financial responsibility regulations. However, while these funds may assist UST owners/operators in complying with EPA's financial responsibility regulations, many funds may not provide the stable, proven protection UST-owning firms need when a leaking UST is discovered. Particularly vulnerable in many states are UST owners/operators with tanks that leaked before the state fund was established.

Table 6. Summary of State Underground Storage Tank Trust Fund Laws as of January 4, 1991

State	Fund Coverage Cleanup/ 3rd Party	Tank Owner/Operator Responsibility or "Deductible" Cleanup	3rd Party	Maximum Fund Limits	Cap or Trigger On/Off	Funding Method	Mandatory Participation	Incentive Program	Loan Program	EPA Approved
Alabama	Both	$5 K/leak	$5 K/leak	$1 M/leak	$7.5 to 10 M	Annual tank fee (100 1990; $10–$150 thereafter)	Yes	No	No	Yes
Alaska	Cleanup only	10% to $25 K	N/A	1 M/leak	None	$50–$500 annual tank fee	Yes	No	No	Not submitted
Arkansas	Both	$25 K/leak	$25 K/leak	$1/leak	$15 M	.2c gal tank fee; $35 annual tank fee	Yes	No	No	Submitted*
Arizona	Cleanup only	$5 K or $25 K Ust owner's option	N/A	$150 K or $250 K, depending on deductible until 12/31/91 then $135 K or $225 K	None	$100 annual tank fee; 1c gal on regulated substances stored	Yes	No	Yes	Submitted*
California	Both	$10 K/leak	To be determined by regulation	$1 M/leak	None	.6c gal annual tank fee	Yes	No	Yes	Resubmitted*

Colorado	Both	$10 K	$25 K	$1 M; $2 aggregate	$4 M	$30 annual tank fee; $25–50 environmental surcharge per tanker of gasoline	Yes	No	No	Submitted*
Connecticut	Both	$10 K	$10 K	$1M	$5 M to $15 M	1% gross receipts tax on petroleum products (not heating fuels) quarterly	Yes	No	No	Rejected—Enabling legislation proposed
Delaware	Both	$100 K	$300 K	$1 M/facility	None	$50 annual tank fee	No	No	No	Not submitted
Florida	Both	Ranges between $500–$25 K; not to exceed $500 for 1st 2 years	Ranges not to exceed $100 K	$1 M/$2 M; aggregate/ per site	$5 M to $50 M unobligated	$50/tank one time fee w/$25/tank annual renewal; 10–20 cents bbl excise tax	Yes	Expired 9/1/89	No	Submitted*
Georgia	Both	$10 K/leak	$10 K/leak	$1 M/leak, up to $2 M aggregate	$10 to $20 M	.1c gal on petroleum products stored	No	No	No	Yes

Table 6. (continued)

State	Fund Coverage Cleanup/ 3rd Party	Tank Owner/Operator Responsibility or "Deductible"		Maximum Fund Limits	Cap or Trigger On/Off	Funding Method	Mandatory Participation	Incentive Program	Loan Program	EPA Approved
		Cleanup	3rd Party							
Idaho	Both	$10 K/leak	$10 K/leak	$1 M/leak, up to $2 M aggregate	$10 M to $20 M	1c gal on all petroleum products paid by first recipient (railroad fuel exempt); $5–$25 annual tank fee	Yes	No	No	Not submitted
Illinois	Both	$10 K/leak; varies based on registration status	$10 K/leak; varies based on registration status	$1 M/leak up to $2 M aggregate	None	.3c gal excise tax	Yes	No	No	Yes
Indiana	Both	$35–$25 K/ leak based on upgrade status	$35K–$25K based on upgrade status	$1 M/leak	$10 M	$290/tank annual fee	Yes	No	No	"Not Formally Submitted"; according to state

Iowa	Cleanup; 3rd party limited to orphans not covered by Federal Insurance program, banks participating in loan program and county foreclosures	25% up to $250 K for leaks discovered after 5/4/89	25% up to $250 K for leaks discovered after 5/4/89	$1 M/leak; up to $2M aggregate	None	$150 annual tank fee; $300 if not upgraded; .85c gal fee on petroleum products	Yes	Yes, for leaks reported after 5/5/89 and before 10/26/90; states pays 75% owner 25%	Yes-loan guarantee program	Yes
Kansas	Both	Marketers with 12 or less tanks $10 K 13–99 tanks- $20 K >99 tanks-$60 K	Not yet established	$1 M/leak; up to $2 M aggregate	$2 M to $5 M	1c gal. fee on petroleum product dist., mf'd or imported	Yes	No	No	Submitted*
Louisiana	Both	$15 K/leak	Same	$1 M/leak	$2 M to $6 M	$13.50/ 9,000 gals of motor fuel loaded at bulk plant	Yes	No	No	Yes

Table 6. (continued)

State	Fund Coverage Cleanup/ 3rd Party	Tank Owner/Operator Responsibility or "Deductible" Cleanup	3rd Party	Maximum Fund Limits	Cap or Trigger On/Off	Funding Method	Mandatory Participation	Incentive Program	Loan Program	EPA Approved
Maine	Both	1 facility, $2.5 K up to $100 K for over 30 facilities	Same	$1 M/leak, $2 M aggregate	None	$35/ complying tank; $130–500 non-complyhing tank; .44c/ bbl fee on gasoline	Yes	No	No	Submitted*
Massachusetts	Both	$5 K-10 K depending on # of tanks	Same	$1 M/leak	$10 M to $30 M	$200 annual tank fee; $50 charge per delivery	Yes	Yes if tank complies	No	Not submitted
Michigan	Both	$10 K/leak	Same but must upgrade w/in 3 years	$1 M/leak	$300 M	.875 gal. fee on petroleum products	Yes	No	Provides interest subsidies to lenders loans	Yes
Minnesota	Both	10%/leak up to $100 K	Same	$1 M/leak	$5 M	1c gal collected at rack	No	No	No	Yes

State	Coverage	Deductible	Deductible (after date)	Per leak	Aggregate	Fee					
Mississippi	Both	0; $5 K/leak after 6/30/92	0; $10 K/leak after 6/30/92	$1 M/leak	$6 M	.2c gal on motor fuels	Yes	Yes, until 6/30/92	No	No	Yes
Missouri	Both; 3rd party: Property damage only	$25 K/leak; 50% of next $25 K; 25% of next $50 K	Same	$1 M	$2 M to $6 M	Annual tank fee of $100	No	No	No		Not submitted
Montana	Both	50% of 1st $35 K; 100% of subsequent costs up to $1 M	Same	$1 M/leak	$4 M to $8 M	1c gal. fee on gasoline distributed 7/1/89-until 6/30/91; 0.75c/gal after 7/1/91	Yes	No	No		Yes
Nebraska	Cleanup only	First $10 K + 25% of next $60 K reported after 7/17/86	None	$1 M/leak	$3 M to $5 M	$25/tank + per gal fee of .3c on gasoline and .1c on other petroleum products	Yes	No	No		Submitted*
Nevada	Both	$25 K/leak $1 K/leak for heating oil tanks	Same	$1 M/leak; 2M/ aggregate	$5 M to $7.5 M	$50/tank annual fee .6c gal fee on gasoline, diesel and home heating oil	Yes	No	No		Not submitted

Table 6. (continued)

State	Fund Coverage Cleanup/ 3rd Party	Tank Owner/Operator Responsibility or "Deductible" Cleanup	3rd Party	Maximum Fund Limits	Cap or Trigger On/Off	Funding Method	Mandatory Participation	Incentive Program	Loan Program	EPA Approved*
New Hampshire	Both	5 K-$30 K depending on no. facilities	Same	$1 M/facility	$10 M	6c gal on gasoline and diesel	Yes	No	No	Submitted*
New Jersey**	State/ Cleanup only	N/A	N/A	None	$50 M	Variable per bbl fee on petroleum	N/A	No	No	No
New Mexico	Cleanup only	First dollar coverage if tank complies	N/A	None	$25 M	$100/tank fee annually; $80 per 8,000 loading fee on gasoline and special fuels	Yes	Yes expires 3/92	No	Yes
New York**	State/ Cleanup only	N/A	N/A	None	$25 M	1c bbl on petroleum	N/A	No	No	No

North Carolina	Both	$50 K/leak	$100 K/leak	$1 M/leak	$5 M to $15 M	Annual tank fee (45–$75/tank depending on tank size) Annual $5.5 M appropriatio to be divided between commercial and non-commercial tank funds	Yes	Expired 10/1/89	No	Yes
North Dakota	Cleanup only	$7.5 K/leak plus 10% of $100 K > the deductible	N/A	$100 K	$1 M to $3 M	.225c gal. on gasoline, kerosene, diesel & heating fuel; annual $25/UST fee; $10/above-ground tanks	Yes	No	No	Submitted*

Table 6. (continued)

State	Fund Coverage Cleanup/3rd Party	Tank Owner/Operator Responsibility or "Deductible" Cleanup	3rd Party	Maximum Fund Limits	Cap or Trigger On/Off	Funding Method	Mandatory Participation	Incentive Program	Loan Program	EPA Approved
Ohio	Both	$50 K/tank; can be revised later; special provisions for owners > 6 tanks; if pay $300 tank fee then 10 K/tank	Same	$1 M/leak; up to $2 M aggregate	$15 M $30 M	$150 annual tank fee through 1990; to be determined in subsequent years; < 6 tanks who pay $300 can reduce deductible	Yes	No	Yes	Yes
Oklahoma	Both	$5 K/tank	Same	Up to $1 M	$10 M	1c gal on motor and diesel fuels and blending materials	Yes	No	No	Yes
Oregon**	State/ Cleanup only	N/A	N/A	None	None	disposal tax on hazardous waste	N/A	No	No	No

State										
Pennsylvania	Both	variable; can be revised no lower than $50 K/tank	variable; can be revised no lower than $100 K/tank	$1 M/leak	None	$50/tank fee annually; Tank indemnification fund fee to be determined by Board	Yes	No	Yes—for owners of 20 tanks/less	Not submitted
South Carolina	Both	$25 K/leak	$25 K/leak;	$1 M/leak	$15 M	$100/tank annual fee; .5c gal environmental fee	Yes	Expired 12/31/89	No	Submitted*
South Dakota	Both	$10 K/leak	Same	$1 M/leak	$2 M to $5 M	"Inspection fee" of 1c/gal on all petroleum products stored	Yes	Yes	Yes—guaranteed loan program for small businesses	Submitted*
Tennessee	Both	$10 K to $50 K, depending on # of tanks	$10 K to $150 K, depending on # of tanks	$1 M/leak	$50 M	$125/tank fee annually	Yes	No	No	Submitted*

Table 6. (continued)

State	Fund Coverage: Cleanup/3rd Party	Tank Owner/Operator Responsibility or "Deductible": Cleanup	3rd Party	Maximum Fund Limits	Cap or Trigger On/Off	Funding Method	Mandatory Participation	Incentive Program	Loan Program	EPA Approved
Texas	Cleanup only for aboveground tanks (ASTs) and USTs	$10 K/leak	None	$1 M/leak	$50 M $125 M	Loading fee collected at rack $12.50–$50 depending on tank size + add'l $25 for each 5,000 gals above 10,000 gals; annual fee of $50/UST; $25/AST	Yes	No	No	Submitted*
Utah	Both; tank and soil tests must be completed prior to coverage	$25 K/leak	Same	$1 M/leak	$12.5 M $17.5 M	$250/tank; after 7/1/93 $150/tank	Yes—for all registered tanks	No	No	Yes

State									
Vermont	Both	10 K/leak	$0 (no deductible)	$1 M/leak; $1 M 3rd party	None	1c gal on gasoline and diesel fuel (collected at rack), varible annual tank fee ($100–200) based on tank size	Yes	No	Yes–for owner/ 20,000 gals or less per facility	Yes
Virginia	Both	$50 K/leak $200 K/ aggregate Deductible under insurance contract, to be determined	$150 K/leak $200 K/ aggregate	$1 M/leak	$10 M $20 M	.2c tax on motor fuels	Yes	No	No	Submitted*
Washington***	Both	Same	Same	State reinsures insurer up to $1 M/ leak; $2 M aggregate	None	Percentage of premium from insurer	Yes	No	No	No
West Virginia	Both	$5 K, $15 K, or $25 K at tank owner's option	Same	$1 M/leak/ $2 M aggregate	$2 M to no limit	Annual fee based on deductible amount and age of tank	No, but all owners pay $100 annual capitalization fee	No	No	Not submitted

Table 6. (continued)

State	Fund Coverage Cleanup/3rd Party	Tank Owner/Operator Responsibility or "Deductible" Cleanup	3rd Party	Maximum Fund Limits	Cap or Trigger On/Off	Funding Method	Mandatory Participation	Incentive Program	Loan Program	EPA Approved*
Wisconsin	Both	$5 K/leak < 7/1/93; $10 K > 7/1/93	Same	100–999 tanks; $1 M/leak; $2 M/annual aggregate all others: $200 K/site	$25 M	Inspection fee of 20¢/50 gal bbl on product imported/inspected in State	Yes	No	No	Submitted*
Wyoming	Both	First dollar coverage if tank complies	$30 K effective 1/1/90	$1 M/leak	$10 M	$200/tank annual fee; self-insured owners $150/tank annual fee; 1¢/gal tax on gasoline	Yes	No	No	Not submitted

Clearly, many UST owners/operators faced with no other options will welcome the coverage provided by any EPA-approved state financial assurance fund — regardless of its limitations. These owners/operators, however, should understand the drawbacks of a particular state fund, as well as the improvements that could make the fund more responsive to his needs.

REFERENCES

1. Anderson, Myra R., State UST financial assurance funds: disasters waiting to happen, Environmental Claims Management, Vol. 1, No. 3, Spring 1989, 313–321.
2. Financial Responsibility For Underground Storage Tanks, The Risk Report, International Risk Management Institute, Inc., October, 1989.
3. U.S. Government Accounting Office, Superfund — Insuring Underground Petroleum Tanks, GAO/RCED-88-39, January 1988.
4. Florida Department of Environmental Regulation, Summary of Testimony Prepared by Marshall T. Mott-Smith, State of Florida Department of Environmental Regulation, Storage Tank Regulation Section, for the Small Business Subcommittee on Antitrust, Impact of Deregulation and Privatization, U.S. House of Representatives, October 31, 1989.
5. Anderson, Myra R. and William P. Gulledge, Do we need to update underground tank regulation? Risk Management, August 1989, 36–41.
6. Thompson, Paul S., Conn, W. David, Geyer, L. Leon, Financial Responsibility Provisions for Underground Storage Tanks in Virginia, Virginia Water Resources Research Center Bull. No. 161, Blacksburg, Virginia, February 1988.
7. U.S. Government Accounting Office, Underground Petroleum Tanks — Owners Ability to Comply with EPA's Financial Responsibility Requirements, GAO/RCED-90-167FS, July 1990.
8. U.S. Environmental Protection Agency, Office of Communications and Public Affairs, Note to Correspondents, April 15, 1991; see also, UST Financial Assurance Deadline Extended 26 Months for Small Firms, *Groundwater Pollution News,* January 7, 1992, p. 5.
9. Faulkner, Barbara, Petroleum Marketers Association of America, memorandum to Association Executives, Motor Fuels Steering Committee, regarding Update on State UST Trust Funds, January 10, 1991.

CHAPTER 11

Lust, Litigation and the Common Law

Leon Geyer

CONTENTS

0-87371-402-4/93/$0.00+$.50
© 1993 by Lewis Publishers

Lust, Litigation, and the Common Law

11.1 INTRODUCTION

Although much environmental law today is the result of legislation, the common law, or judge-made law, has in this country been the historical source of law providing the remedy for environmental damage inflicted by one party upon another. The common law comprises the body of those principles and rules of action relating to persons and property which derive their authority solely from usage and customs or from the judgments and decrees of courts recognizing, affirming, and enforcing such usage and customs. Common law is independent of statutory rules. However, if there is a conflict with statutory language (legislature-made rules), then the statutory language determines the outcome.

Because the statutory environmental law has not completely supplanted the common law, we must review aspects of the common law which impact upon the owner, operator, or lessor of underground storage tanks (UST). Although federal and state law establishes a regulatory scheme for USTs, state common law rules apply to individual situations.

The general principles of common law which might be applied to UST problems will be discussed below. Case law has not been fully developed in all states, but certain common law principles have been developed and vary only slightly from state to state in interpretation and application.

Historically, the function of the common law has been to redress civil wrongs between private parties. Corrective justice has been accomplished by holding a person liable who caused or allowed to be caused injury to the person or property of another person as a result of a breach of a legal duty. Most of these duties arise from the concept of tort—a private or civil wrong or injury independent of contract. The duty is imposed by common or statutory law. There must be a violation of a duty owing to one person (plaintiff) by another (defendant). The three elements of every tort are: existence of a legal duty from one person to another, a breach of the duty, and a resulting injury or damage. If actionable under tort law, environmental injuries to a person or to property are remedied by money damages or, in some cases, by injunctive relief.

The advent of environmental statutory law was the result of a failure of the common law to address community-wide disruption (public wrong) typical of environmental degradation and the advance of scientific knowledge. Although the common law courts focused on damages and not future problems, early environmental statutes were almost all prospective and redressed public rather than private wrongs. Many were health-based and not environmental per se.

Since the 1970s, federal and state legislatures have enacted broader environmental protection measures and the common law now provides private damages in some cases and fills in gaps remaining in the statutes.

Beyond the scope of this chapter but important to resolving the role of the common law in environmental litigation is whether the advent of "comprehensive" state and federal environmental regulatory schemes preempt some or all common law remedies. Although legislatures generally have the right to preclude common law remedies, each individual piece of legislation must be reviewed to determine its impact on common law rights.[1]

The function of tort law is to force those who generate external costs to bear those costs as a means of redressing individual harm and as a means of moving toward a more efficient allocation of resources.[2] External costs are those not internalized in the production of a good or service. For example, the cost of gasoline is internalized to the seller of gasoline. The vapors are externalized into the atmosphere and the "cost" of pollution is borne by consumers of air rather than purchasers of the retail product.

Common law relating to leaking underground storage tanks (LUST) developed with the advent of turn of the century fire codes and the industrial practice of placing gasoline storage tanks underground. Following the automobile into suburbia and beyond, lawsuits increased with tank leaks in residential areas. Although the UST area is now extensively regulated by statute and regulation, traditional common law remedies remain available for victims of environmentally damaging leaks. This applies both to leaking tanks governed under Resource Conservation and Recovery Act (RCRA)[3] and tanks exempt from RCRA technical and financial requirements. State environmental common law generally recognizes one or more of the following theories of recovery for wrongs: nuisance, trespass, negligence, or strict liability (in some states, strict liability only for "ultra-hazardous activities").

11.2 NEGLIGENCE

Negligence is defined by the Restatement of Torts (2nd) as conduct which falls below the standard established by law for the protection of others against unreasonable risk of harm. *Conduct*, not conditions, means that a successful negligence action requires that the defendant must be charged with some act or failure to act. It must be shown that the defendant was under a duty to conform to a standard of conduct, that defendant breached the established duty, that there was a reasonably close connection between conduct and the resulting injury (proximate cause of injury), and that plaintiff suffered actual loss or damage. The duty of care, or a duty to conform to a reasonable standard, which may be imposed by statute, regulation, earlier judicial decisions or the jury is "the standard of conduct of an ordinary, reasonable, and prudent person under the given conditions and circumstances." Examples of jury questions are: How often would an ordinary, reasonable and prudent UST operator check his/her tanks? What type of records would be kept? Did defendant violate this duty?

11.2.1 Duty to Use Reasonable Care

In Murchie v. Standard Oil Company,[4] plaintiff alleged the oil supplier was guilty of negligence in carelessly, recklessly, and negligently delivering excessive amounts of oil. Homeowner argued that the supplier knew, or by the exercise of ordinary care and prudence should have known, that there was a leak. Home owner had a 3000-gal fuel tank. In 2 prior years, home owner had used less than 3000 gal per year. In the two years of the alleged negligence, the home owner averaged 13,000 gal per year. Defendant supplier had a contract with the home owner to "keep the tank full."[5] The court ruled that it was a jury question as to whether or not the defendant supplier was negligent. The jury could infer that the defendant breached a duty.

In another case, plaintiff alleged a nuisance and continuous trespass. Such actions give rise to a cause of action as long as the trespass or nuisance continues. The court held that because defendants did not have exclusive use and enjoyment of property, the prior owner ceased to be liable for negligence when the ownership or possession and control passed to another.[6]

> Even where a continuous trespass or nuisance exists, liability of the owner terminates after conveyance at such time as the new owner has had reasonable opportunity to discover the condition by making prompt inspection and necessary repairs.[7]

While this was a valid defense for a prior owner/defendant to a leak that continued to occur after sale of property, the case implies that a new purchaser of property has a duty to inspect and make sure his premises are free of LUST. In accord is Restatement of the Law Torts 2d Sections 351, 352, which state that when a prior landowner ceases to be either an owner or an occupier, and his responsibility ends for the subsequent developments, whether natural or artificial, which existed at the time the successor owner took possession. An exception to this rule is when an undisclosed dangerous condition is know to the prior seller.[8]

A cause of action was sustained against a supplier of gasoline where the supplier failed to maintain, inspect, test or otherwise monitor the tanks.[9] The court said that the supplier knew or should have known that there was an underground loss of gasoline occasioned by a leak. Liability may be charged to the supplier if the supplier had notice, actual or constructive, of the dangerous condition.[10] A supplier has a defined duty of care under the common law.

In Cooper et al. v. Whiting Oil Company, Inc.,[11] the service station owner and the neighboring landowners brought an action against the oil supplier for alleged negligence in maintaining gasoline storage tanks at the station. After being notified of a potential leak in a tank, the oil supplier failed for over a week to check out the tank. The tank in question was installed by the supplier when plaintiff/service station operator began to operate the service station. The tank was never inspected, repaired or replaced by the supplier. In classic common-law tort language, the court stated that "[T]he burden rested upon the landowner to prove by a preponderance of the evidence that [supplier] was negligent and that the negligence was a proximate cause of property damages alleged."[12] The fact that the gasoline leaked into the adjoining landowner's well does not establish negligence on the part of supplier. The landowners offered no evidence of the supplier's duty to inspect and maintain the tank. No evidence was introduced to show that: (1) company policy or industry practice required periodic inspection; (2) how leaks could be prevented; or (3) how leaks could be detected by appropriate inspection or maintenance procedures.[13] Plaintiff failed to show negligence on the part of supplier prior to notification of the leak or possible leak. [Modern RCRA standards would likely change this common law interpretation or at least supply plaintiffs a standard of care.]

A second interesting problem in this case was a recognition that the supplier, although not proven initially negligent, may have been negligent for failing to act in a timely and proper method after notification of the leak.[14] When damage occurs as a result of more than one cause (prenotification of leak or possible leak and postnotification of leak), it is for the jury to determine what part is attributable to which defendant or in this case, how much the defendant would be liable for postnotification damages. However, plaintiff failed to produce evidence to show with a reasonable degree of certainty the share of prenotice and postnotice damages.[15]

In VanVooren v. John E. Fogarty Memorial Hospital,[16] supplier repaired a leak in an oil line. Plaintiff argued that the supplier was negligent because the supplier failed to replace the soil that was saturated with oil as a "reasonable person" would have done. Plaintiff argued that a jury could have found this failure to replace the soil a negligent act. The court ruled that the defendant supplier owed no duty to adjoining landowner to replace saturated soil.[17] Supplier was not negligent in repairing the leak nor responsible for the leak in the first place.[18] The Court held that although a second defendant, after being advised of the leak and contamination of the plaintiff's water supply, fixed the leak, it was a jury question as to whether the supplier/defendant owed the plaintiff a duty not only to repair the leak, but also to exercise reasonable care that the oil would not contaminate the plaintiff's water supply[19] by removal of the soil.

In Lerro v. Thomas Wynne, Inc.,[20] the Court held that the owner and operator of an apartment house was negligent in (1) failing to inspect the oil tank at reasonable intervals and (2) failing to detect a loss of large quantities of fuel oil. Periodic inspections were held to be simple and inexpensive.[21] "Ordinary men know that large quantities of oil soaking into the ground, if uncontrolled, flow in unpredictable directions and therefore involve a risk of seriously affecting other properties."[22]

Although no absolute standard can be fixed by law for the degree of care imposed upon a person dealing with gasoline, a higher degree of care can be imposed on persons dealing with such dangerous substances than required of a person in ordinary circumstances.[23] As one of the elements of negligence, the plaintiff would allege [and must prove by the preponderance of the evidence] that a defendant did violate a duty of reasonable care in the installation and use of an UST. This task is made easier now that EPA and state regulatory programs elevate the duty of care for tank owners and operators. The governmental standards could be the basis for actions grounded on negligence per se (violation of statute). The trend of the common law negligence cases is to make negligence easier to prove.

Plaintiffs may allege that defendant has violated the following duties:

1. Failure to repair the tanks
2. Failure to clean up a spill
3. Failure to perform on a contractual obligation to a lessee
4. Failure to test for leaks
5. Failure to discover a leak through inventory analysis or mechanical methods within a reasonable time; failure to monitor and discover leak as soon as possible
6. Failure to act based on knowledge of the tank's condition
7. Failure to use reasonable care in construction of the tank — failure to comply with governmental standards
8. Failure to act on a contractual duty to inspect or maintain equipment
9. Failure to warn potentially affected individuals when leak or possible leak is discovered
10. Design failure on part of the manufacturer
11. Design failure or implementation of tank siting by installation contractor/ owner/others
12. Use of unprotected or inadequately protected tanks; the new UST regulation will raise the negligence standard
13. Failure to prevent the leak of gasoline by the use of appropriate technology
14. Failure to warn those who may be injured as a result of the leak
15. Failure to remedy the leak and clean up the spill as soon as reasonably possible[24]
16. Failure to implement tank integrity programs including the removal and replacement of faulty systems in a timely manner
17. Failure to properly install the tank

11.2.2 Duty to Mitigate

It is important to reduce damages by mitigation. When the leak and the resulting damages continue, the statute of limitations does not begin to run until the conduct causing the damage is abated or stopped.[25] The plaintiff did not have a duty to mitigate where the suggested "repair" of the problem was inadequate.[26] Plaintiff/lessees were not required to have done better in finding the leak in the tanks than the defendant/ lessors whose experts failed to reveal the causes of the losses.[27] Failure to act by owner within 3 days after discovery of leak was held to violate the duty to mitigate in a timely manner.[28]

In Larrance Tank Corporation v. Burrough,[29] the court ruled that it was a jury question as to whether plaintiff notified defendant tank manufacturer within a "reasonable time" after the plaintiff "should have discovered" the defect in the tank. Plaintiff does have a duty to mitigate damages.

11.3 NUISANCE

A private nuisance is an unreasonable interference with plaintiff's use and enjoyment of their land. The concept is that defendant's use or misuse of their land interferes with another's use of his or her land. The environmental interference with use and enjoyment may include odors, air pollution, water pollution, noise, and contamination of subsurface waters or land. The maintenance of a hazardous condition may give rise to a nuisance action. A private cause of action is primarily designed to secure injunctive relief (court order to stop activity) for the future or damages for past interference.

In Kostyal v. Cass[30] the question of whether a nuisance existed was one for the jury. The plaintiff alleged damage due to a leak in defendant's gasoline tank. The defendant was a public authority. The court laid out three types of actions in nuisance which can be maintained against a public authority:[31]

> (1) nuisances which result from conduct of the public authority in violation of some statutory enactment; (2) nuisances which are intentional in the sense that the creator of them intended to bring about the conditions which can be found as a fact to constitute a nuisance; and (3) nuisances which have their origin in negligence, that is, in the failure of the creator of the conditions constituting a nuisance to exercise due care.

The question of whether a nuisance exists is a factual decision for the jury.

While nuisance and negligence are different in their nature and consequences, these torts may be and are frequently coexisting and practically inseparable. Acts and failure to act constituting negligence often give rise to a nuisance in the case of LUSTs. A nuisance does damage to another or interferes with the free use, possession, or enjoyment of, or physical and safe occupation of another's property.

Fuel storage tanks do not constitute a nuisance per se but may become a nuisance by reason of the manner in which they are constructed or operated.[32] Nuisance per se is in reality another term for strict liability.[33] Unlike other torts, nuisance is not concerned with the nature of the conduct causing the damage, but the relative importance of the interests interfered with or invaded. The essential element is that persons have suffered harm they ought not to bear.[34]

In Yomers v. McKenzie,[35] the court reaffirmed that the establishment of a gasoline filling station (and its underground tanks) does not constitute a nuisance per se. But, a station (and its leaking tanks) may become a nuisance because of its location or manner of operation. Mel Foster Co. Properties Inc. v. American Oil Company[36] affirms that gasoline leakage is an actionable nuisance. The case also affirms the rule that a nuisance is not abated until the use and enjoyment of the property is no longer interfered with. A LUST is a temporary nuisance until the pollution is discovered and abated. The nuisance is also permanent in that it damages the ground, water, and other natural elements.

> An attempt to classify chemical pollution as a permanent or temporary nuisance is further complicated by the presence of rapidly changing scientific technology. Scientific knowledge enables society to successfully clean up pollution once thought to be permanent; it also reveals hidden dangers in chemicals once thought to be safe...The terms "permanent" and "temporary" are somewhat nebulous in that they have practical meaning only in relation to particular fact situations and can change in characterization from one set of facts to another.[37]

Permanent damages would be appropriate in cases where the damages do not fit neatly into a traditional box of temporary or permanent. Even though oil contamination may eventually leave the ground, an award of permanent damages is appropriate.[38]

11.4 STRICT LIABILITY

Strict liability for injury due to environmental damage is imposed without regard to fault generally because of the dangerous nature of an activity. Strict liability is imposed regardless of fault for ultrahazardous or dangerous activities. Not all states impose strict liability on hazardous activities. The courts rely on different formulations to determine what activities should be covered by the principle of strict liability and courts apply the same formulation differently. The Restatement of Torts (2nd) Section 520 outlines the factors that are used in determining whether an activity is abnormally dangerous:[39]

> In determining whether an activity is abnormally dangerous, the following factors are to be considered:
>   (a) existence of a high degree of risk and some harm to the person, land or chattels of others;
>   (b) likelihood that the harm that results from it will be great;
>   (c) inability to eliminate the risk by the exercise of reasonable care;
>   (d) extent to which the activity is not a matter of common usage;
>   (e) inappropriateness of the activity to the place where it is carried out; and
>   (f) extent to which its value to the community is outweighed by its dangerous attributes.

Strict liability may be imposed on two basic theories — "non-natural land use" or "abnormally dangerous activity." Under the non-natural land use theory, strict liability is imposed upon someone who maintains a potential hazard which "escapes" and causes injury. Strict liability has been imposed upon the owner or occupier of land in several jurisdictions[40] but the application to UST has been inconsistent. Not all courts or juries would view the storage of gasoline as a non-natural use of the land. It may also be difficult to convince a court that the storage of gasoline is an abnormally dangerous activity.[41] Strict liability may be imposed on the tank manufacturer for injuries from a product that is in a defective condition and is unreasonably dangerous based on principles of products liability.[42]

The court in Yomers v. McKenzie[43] adopted a strict liability rule for the storage of large quantities of gasoline immediately adjacent to private residences with wells. The LUST owner was not using the land in the common and natural way. The LUST owner/land occupier had artificially produced the potential danger. The operation of the USTs was considered an "abnormally dangerous activity" involving a "high degree of risk of some harm to the person, land, or chattels of others."[44] The risk of harm cannot be eliminated by the exercise of reasonable care. The doctrine of strict liability was imposed in this case because the activity is not of "common usage" in the neighborhood. Thus, the plaintiff did not need to show or prove defendant LUST owner's negligence.[45]

In United States v. White Fuel Corporation,[46] oil from defendant's farm tank seeped into a harbor. The court held defendant liable for noncompliance with the Refuse Act.[47] The undisputed fact that the oil owned by the defendant leaked from his property was sufficient to sustain a judgment against defendant in absence of proof that a third party was responsible. Indirect percolation rather than direct flow was immaterial.[48] In upholding the conviction, the court specifically rejected any requirement of intent or willfulness on the part of defendant. The dominant purpose of the Act is to require people to exercise whatever diligence they must to keep refuse out of public waters. "Due care" is not a defense that would allow a polluter to avoid judgment on the grounds that he took precautions or conformed to industry-wide or commonly accepted standards.[49] But, the court stated

> ... it remains true that the standard of proof of exculpatory facts is of a much more rigorous and exceptional nature than that of reasonable care. Particularly is this true of an enterprise such as that we deal with here — an oil tank farm, with its inherent risk of spills, leaks, and seepage, abutting navigable waters. Such a high risk enterprise carries with it, even under conventional tort law, a high burden of responsibility.

This case imposed strict liability on the defendant as an interpretation of a federal statute. The case also provided a set of defenses to a strict liability case.[50]

In Southern Co. Inc. v. Graham,[51] the court upheld a finding of strict liability for the installation of a defective tank. After seller installed tank, pumps, and lines, buyer and customers smelled gas in the water supply. The State had adopted the doctrine of strict liability in torts in products liability. The doctrine does away with the necessity of proving negligence to recover, but the plaintiff must still "show that circumstances surrounding the transaction were such as to justify a reasonable inference of the probability rather than a mere possibility" that the defendant's product is responsible for the problem.[52] The trial court could find that the defendant/tank installer was liable for the subsequent leak under a strict liability standard.

In Branch v. Western Petroleum, Inc.[53] the court upheld the application of the Rylands v. Fletcher[54] standard of strict liability for percolating oil from a pond that leaked into plaintiff's well. Holding defendant liable for pollution, the court said the ponding of water in an area adjacent to plaintiff's wells constituted an abnormally dangerous and inappropriate use of the land.[55]

Applying Louisiana law, the court imposed liability on the gasoline distributor for damages sustained on underground cables owned by the plaintiff/ telephone company.[56] The distributor was responsible for the maintenance and upkeep of the tanks and lines which leaked and caused the damage. Article 2317 of the Civil Code imposes liability (strict) for "things which we have in our custody."[57] The guardian of the leaking tanks is liable for his legal fault in maintaining the defective gasoline tanks.

In accord is City of Northglenn, Colorado v. Chevron USA, Inc.[58] which held that, under Colorado law, strict liability for the storage of gasoline in a residential neighborhood could be imposed by a jury. The implication is that the storage of gasoline is an abnormally dangerous activity.

In Hudson v. Peavey Oil Co.,[59] the court held that the storage of gasoline in conjunction with the operation of a service station is not an ultrahazardous or abnormally dangerous activity. Seepage can be eliminated by the exercise of reasonable care and therefore the facts do not fit the legal requirements for imposition of strict liability for LUST.[60]

11.5 TRESPASS

Trespass requires an interference with plaintiff's possessory interest in land. An entry upon plaintiff's land may be by the defendant or by some physical or observable object which defendant has caused or allowed to enter plaintiffs property. This cause of action has potential for recovery by plaintiffs in cases of physical above-ground water contamination or contamination of percolating groundwater and soil. Environmental type trespass cases have had limitations imposed by the courts which include:

1. Plaintiff must be in possession of property.
2. The injury must be related to an invasion of property.
3. Most states have moved to an intentional, negligent, or hazardous activity standard and away from the strict liability standard for trespass based on invasion of a plaintiff's property. Plaintiff may not recover for inadvertent invasion by defendant without a showing of negligence or strict liability on the part of the defendant.

4. Some jurisdictions require "visible" or "tangible" matter from pollutants vs. "scientific evidence" of invasion.
5. The American water rule followed in most states is that a land owner is not absolutely entitled to pure percolating ground water. Negligence, defendants knowledge of likely contamination or ultrahazardous activities must be shown by:
   a. Existence of a high degree of risk and some harm to the person, land or chattels of others
   b. Likelihood that the harm that results from it will be great
   c. Inability to eliminate the risk by the exercise of reasonable care
   d. Extent to which the activity is not a matter of common usage
   e. Inappropriateness of the activity to the place where it is carried out
   f. Extent to which its value to the community is outweighed by its dangerous attributes

Common law trespass cases have been brought against defendants arising from alleged underground travel from defendant's land into plaintiff's land and water wells. The leaking substances are an "unlawful intrusion" on plaintiff's property. However, some courts have held that "trespass is an intentional harm"[61] and the trespasser must intend the act which amounts to or produces the unlawful invasion. The invasion must represent willfulness on the part of defendant. In the absence of knowledge or notice, the defendant is not likely to be found liable for trespass of noxious fluids.[62] A cause of action in trespass (and negligence) might arise where defendant has been notified of a leak or potential leak and defendant fails to act to abate the leak.

Trespass may be found for unintentional trespass if it arises out of defendant's negligence or the carrying on of an ultrahazardous activity.[63] Successful trespass cases often require proving negligence or the adoption by the state of the strict liability standard.

11.6 FEDERAL COMMON LAW

Although most common law cases have arisen under state law due to the proximity of the land polluted by an oil leak, at least one case provided for relief under the federal common law[64] of nuisance.[65] The application of federal common law of nuisance, like state law, is one for the jury as to whether substantial evidence exists to support a violation of the common law. The federal common law of environmental pollution has been preempted in some cases by federal statute.[66] Where there is interstate damage of land, federal common law would still be applicable.

11.7 SOURCE OF LEAK AND CAUSATION

Proving the source of a gasoline leak is a factual and sometimes difficult task. In Yomers v. McKenzie,[67] the court found that the jury could find factually where the leaks came from when confronted with the following options: the plaintiff's well was within 60 to 70 feet of defendant's 1000 gal tank; the tank had been in place for 20 years; the depth of the well was below the tank; there was evidence of seepage of gasoline around the tank; the soil of the excavation smelled of gas; a test of the tank after removal developed no evidence of a leak; a second and smaller tank on defendant's property was tested with negative results; a 1500-gal tank truck had spilled gas within a mile of the plaintiff's home; and, there was evidence of seepage of gasoline and kerosene on the ground at another filling station some 300 ft from plaintiff and on the other side of the road. Based on these "facts", the court stated, "The gasoline had to come from somewhere."[68] The jury could make a finding of fact or inference based on this "proof" of pollution contamination. A reading of cases indicates the importance of factual information upon which the jury can base its decision.

In A. H. Grove and Sons, Inc. v. Commonwealth of Pennsylvania,[69] the court held that circumstantial evidence was substantial and competent to support the state's finding that the defendant's property was the probable source of contamination. The operator was required to bear the cost of testing to discover how to abate the problem.[70] This circumstantial evidence can lead to common law or statutory liability. Likewise, the cost of securing cleanup methods can be charged to the polluter under statutory authority.

Causation of damages to a plaintiff by an UST owner/operator, an essential element of proximate cause, must be established by the evidence. The "fact" that a gasoline leak "caused" the damage can be proven by circumstantial evidence.[71] A series of inferences can be used, but competent and probative evidence must be offered to the jury or trier of fact. Plaintiffs have lost cases because they could not prove the damages were from the defendant's tanks.

The plaintiff must prove by the preponderance of the evidence that an alleged LUST owner/operator was negligent and that the leak was the proximate cause of the plaintiff's damages. If damages occur as a result of more than one cause, it is for the jury to apportion the damages. In at least one case,[72] plaintiff's case failed because he/she failed to establish the causal relationship.

11.8 DAMAGES

The court has held that one measure of damages may be the difference in the value of the real estate before the wells were polluted by the gasoline and the value thereafter.[73] This was supported by testimony that the wells were not clean after 5 to 6 years. Failure by plaintiff to show records of the difference between quantity of gasoline purchased and sold led a court to rule that "lost profits" were speculative.[74] Excessive cost of well installation was denied where similar wells were installed for one third of plaintiff's installation. Plaintiffs must prove actual damages.

In Moore v. Mobil Oil Co.[75] the court held that a lower court's order requiring that the defendant implement measures to assure that the gasoline "already in the soil and groundwater" did not continue to pollute the plaintiff's wells was improper. The appellate court said that the evidence in the case did not show that a workable and scientifically proven solution to the problem of continued subsurface seepage of gasoline in the ground could be found.[76] The trial court's solution was not equitable. Money damages for actual expenses in securing potable water and inconvenience in securing the water supply both before and after the date of hearings were proper, as were loss of value of property and awards based on future inconveniences.[77] The dissent in the above case argued persuasively for an injunction.[78] With the advent of technology in LUST cleanup, such remedies may be appropriate in the future.

In a case for damages to adjacent swamp land from a leaking pit, the court held liable the company responsible for pit construction and one company who placed toxic chemicals in the pit after they should have known the pit was leaking.[79] This case reminds us that courts will hold that construction and installation of USTs must be done in a worker-like manner and that liability will be extended to "those who know or should have known." Other companies were not liable for activities[80] related to the leak prior to their knowledge of the leak.

In assessing damages in a toxic leak case, the court in Ewell v. Petro Processors of Louisiana, Inc. said,

> We think that the proper measure of damages must be determined from the circumstances of each case, considering such factors as the extent of the damage; use to which the property may be put; extent of economic loss, both as to value and income; and the cost and practicability of restoration.[81]

The value of the property in question was $375 per acre or slightly over $200,000 for 550 acres affected. The restoration of the property according to one witness would cost 170 million dollars (7 years of 100 trucks running continuously).[82] The court was of the opinion that the proper measure of damages is the diminution in value of the affected property and held the jury's finding of no value left in the property was proper.[83]

In Bean v. Sears, Roebuck and Company[84] the court held that it was appropriate for the jury to used the standard of "what it will fetch"[85] before and after the diminution of value. The jury was not required to use as the measure of damage the cost of restoration of property to its former condition, where such a measure of damages was not readily ascertainable.

In addition to the loss of property value, plaintiffs are often accorded reasonable expenses in preventing, reducing, or abating the results of the LUST operator.[86] Defendants may also be liable for the action of a negligent contract or if the defendant LUST owner retains the right and power of directing, in detail, how the work should be done.[87]

Litigation an be extensive and expensive. In Exxon Corp. v. Yarema,[88] 27 litigants and 90 claims were involved. The jury awarded over $1,000,000 in punitive damages. Implied and not actual malice is the proper standard for a finding of punitive damages in a tort action.[89] The court held that a jury could find implied malice where defendant became aware of the leak, delayed the removal of the gasoline from the tanks, and in fact kept refilling them with gasoline. In addition, for 2 years, defendant/LUST owner made no effort to recover the lost gasoline.[90] In addition, the defendant did not inform plaintiff that defendant had lost product but rather speculated that plaintiff's contaminations were from another unlikely source. The court held that a jury could imply that defendant had disregarded the risk of groundwater contamination by its actions and thus be liable for punitive damages.

Damages may also include business interruption by an adjacent business interest. Toxicity of leaked chemicals — acute and chronic — may be sources of damages for LUST. Proof of causation may be difficult for many reasons including quantification of exposure and multiple sources of exposure.[91] Recovery of damages for mental distress is another potential source of litigation concern.

Real property damages may include loss of trees and shrubs, growing crops, rental value of land, reduction in value of land, cost of obtaining potable water (short and long term), cost of soil decontamination, cost of investigating the leak, and cost of technical experts and lawyers. Personal damage may include medical expenses, loss of past and future earnings, impairment to enjoyment of life, interest cost, and punitive or exemplary damages for malicious or reckless conduct.

Current legislation in some cases expands upon the concept of damages to include the total value of damages caused to the environment and/natural resources upon a violation of antipollution laws and not just the diminution in value of the property.[92] These environmental costs are based on statutory authority. Although no common law right of action exists at this stage for environmental damages from LUST, governmental statutes may provide for the recovery of such expenses.[93] Citizen suits to enforce statutory requirements are allowed in some cases. Compensation for the destruction of natural resources are the remedies that legislators are likely to use as a deterrent in the future.

11.9 LUST AND GOVERNMENTAL LITIGATION

11.9.1 Liability of Governmental Units

In Bousquet v. Commonwealth,[94] the court held that the Commonwealth or State (city and counties, too) did not have sovereign immunity from liability where it created or maintained a private nuisance. The Commonwealth had allowed oil from fuel tanks at a state school to seep into a stream that ran through plaintiff's property.

11.9.2 Statutory Rights

Neither federal,[95] state,[96] or local statutes expressly establish a private right of action nor is it likely that the courts will interpret the statutes to imply one.[97]

Federal, state, and local statutes may, however, place a higher standard of care upon the UST owner/operator in common law litigation. EPA regulations (and state programs) on the detection, prevention, and corrections of leaks could amount to negligence per se.[98] EPA requirements include the following:

(1) requirements for maintaining a leak detection system, an inventory control system together with tank testing, or a comparable system or method designed to identify releases in a manner consistent with the protection of human health and the environment;

(2) requirements for maintaining records of any monitoring or leak detection system or inventory control system or tank testing or comparable system;

(3) requirements for reporting of releases and corrective action taken in response to a release from an underground storage tank;

(4) requirements for taking corrective action in response to a release from an underground storage tank;

(5) requirements for the closure of tanks to prevent future releases of regulated substances into the environment; and

(6) requirements for maintaining evidence of financial responsibility for taking corrective action and compensating third parties for bodily injury and property damage caused by sudden and non-sudden accidental releases arising from operating an underground storage tank.[99]

11.9.3 Authority of the State

Courts have supported the constitutionality of regulatory schemes such as RCRA where the state has "decided upon the destruction of one class of property [without compensation] in order to save another which in the judgment of the legislature, is of greater value to the public."[100] Requiring the landowner to abate the leaking storage tank to preserve the value of the neighboring property is a legitimate exercise of the state police power.[101]

In the Matter of Contempt Order (Anderson),[102] the Department of Environmental Quality sought to enforce discovery concerning matters relating to USTs. Although the issue is technical and beyond the scope of this book, it outlines what the authors perceive to be a trend — future litigation is likely to come about by the enforcement of statutory regulations by governmental agencies and not just the neighboring landowner.

11.10 COMMON LAW DEFENSE TO LUST CASES

All common law torts outlined above include problems with proving and establishing amount of damages from an environmental degradation such as a

LUST. Defendants in common law tort litigation may generally avail themselves to several defenses including contributory negligence (approximately 14 states), comparative negligence (approximately 36 states), assumption of risk, statute of limitations, and defense to strict liability.

11.10.1 Contributory Negligence

Even if the defendant is negligent, a "negligent" plaintiff might be barred from recovery due to plaintiff's own action or negligence. Under the concept of contributory negligence, if a plaintiff fails to exercise due care for his or her own safety and such failure is a contributing cause of plaintiff's injury, then the plaintiff will be barred from recovery regardless of the defendant's negligence. The standard for plaintiff's conduct is that of "an ordinary, reasonable and prudent person." It is a factual question for the jury to determine.

Contributory negligence is not a defense to a nuisance action nor to an action based on strict liability.[103] Although plaintiff failed to notify LUST owner, the court ruled that he could not be declared as a matter of law guilty of contributory negligence.[104] It was a question for the jury[105] as to whether or not the plaintiff assumed the risk of a leaking tank as a lessee of UST owner.

11.10.2 Comparative Negligence

Comparative negligence (not allowed in state with contributory negligence) allows for an apportionment of damages between plaintiff and defendant according to fault of each. For example, if plaintiff's damage was $100,000 and the plaintiff was 30% at fault, the reward would be reduced to $70,000. About 36 states have adopted the comparative negligence standard by statute or by case law. Such a defense could be used to reduce damages where the owner was held liable for leaking tanks, but the lessee of the tanks was also negligent for failing to discover the leak in a timely manner or notifying tank owner upon discovery of the leak.

11.10.3 Assumption of the Risk

A plaintiff can be held to have "assumed the risk" if he voluntarily subjects himself to a known risk. The standard for assumption of the risk doctrine is determined by the plaintiff's age, experience, and knowledge. A person must have the ability to understand and consent to the danger involved in the situation. These defenses would generally not be available to a defendant who is the target of a tort suit for damages from a LUST.

11.10.4 Statute of Limitations

Generally, the statute of limitations date for damage to adjacent land runs from the time when the damage becomes apparent and the injured party discovers (or should have discovered) who or what caused it. When damaging conduct continues, the statute of limitation runs from the last date of the last harmful act.[106] Thus, the court may be required to determine when the "damage was consummated" to determine when the statute of limitations has run its course.[107] The date starts when the wrongful conduct causing the damage ends.

In New York Telephone Company v. Mobil Oil Corporation,[108] the Court held that the 3 year statute of limitations applicable to property damage actions barred a complaint alleging improper installation and maintenance of USTs and inadequate and improper closing and sealing of an abandoned tank.

A statute of limitations defense was applied to limit an action for damages in negligence where plaintiff waited more than 3 years[109] to bring the claim.[110] But the 3-year statute of limitations did not bar a cause of action in the same case for a "renewing" nuisance or a "renewing" rather than continuing trespass.[111] Plaintiff was denied a cause of action under the state "Oil Pollution and Hazardous Substances Control Act of 1978" because plaintiff failed to bring the action within 3 years.[112]

11.10.5 Defenses to Strict Liability

Defenses to strict liability would include the intervention by a third party creating the discharge. The North Carolina court gave these examples of defenses to strict liability:[113]

1. Another's oil percolated through a defendant's soil
2. Accidental intrusion
3. Independent contractor caused the leak or negligence of third parties
4. An Act of God or extreme natural disaster
5. An act of sabotage or thievery

11.11 POTENTIAL PLAINTIFF IN A LUST SUIT

Although the majority of LUST cases have been brought as damage claims by adjacent homeowners,[114] lessee's, individuals injured from explosion due to leaking gas (customers), consumers of contaminated water including downstream users and those concerned about the risk of explosion, and those affected by emotional anxiety may be plaintiffs.

11.12 POTENTIAL DEFENDANTS IN A LUST SUIT

Various legal theories have allowed or may allow recovery against the following potential defendants:

1. Station/business owner — Landowner/station owner who leases the premises
2. Station/business operator — The actual operator of the tanks whether the product is for resale or used on premises for the business
3. Refiner — Refiners may have built/leased/own the local station or supply directly to large business consumers
4. Jobber or dealer — Organizations that deliver the gasoline or fuel oil to the station or end user business concern
5. Tank manufacturer — Products liability
6. Seller/installer of the tank

11.13 WHAT TO DO WHEN THE LEAK IS DISCOVERED

At common law, a tortious act is ordinarily covered by the "ordinary, reasonable and prudent person" standard. Modern regulatory statutes require notice and remediation. Complying with common sense and the statutory requirements can reduce the amount of and the potential for common law liability. Juries are less likely to find negligence and neighbors are less likely to bring an action if prompt, courteous, and careful action is taken.[115] The following ideas have been developed as a guide to reducing exposure for LUSTs.

1. Comply with the statutory requirements.[116]
2. Have a leak contingency plan.
   a. Inform employees what to do and whom to contact.
   b. Have preselected cleanup contractor(s) or consultant.
   c. Provide a list of individuals and agencies to contact.
3. Control the leak if at all possible.
4. Write a narrative of steps undertaken to solve the problem.
5. Arrange to have photographs taken.
6. Notify the insurance carrier.
7. Take statements from employees, others about what has happened and what you have done in a timely fashion to resolve the leak.
8. Prepare a statement for the press in case it becomes an issue in the local press.
9. Advise employees who will be the spokespersons to handle inquiries and issue information.
10. Notify any potentially impacted neighbors. Public relations (that I care and I am solving the problem) can be a best line of defense.
11. Notify the landlord if the property is leased.
12. Have a crisis team in place.
13. Draft written reports as soon as possible and submit to authorities as required by law.
14. Consult your attorney.
15. Prepare a site information sheet to include the following:
    a. Location, size, and use of tank
    b. Quantity and duration of leak
    c. How you learned that the tank was leaking
    d. Indication of discharge to soil, groundwater, or water courses
    e. Any unique environmental or social concerns

11.14 WHAT QUESTIONS THE ATTORNEY SHOULD ASK IN A LEAKING STORAGE TANK CASE

Because of the concerns over statutory and common law liability, most tank owners will want to secure legal advice. The lawyer is likely to ascertain the following type of information.

1. Factual information about the leak.
   a. Location, substance, quantity , duration
   b. Human and environmental damage
   c. Indication of discharge to ground, surface or stream water, or soil
   d. RCRA status of facility
   e. Age of tank
   f. Registration status of tank
   g. Location of the leak
   h. Information on the installation of the tank
   i. Warranties by installers or manufacturers of the tank
   j. Records of systematic testing for leaks
   k. History of ownership and use of the tank
   l. Copy of your contingency plan
   m. Information related to your activities since the leak was discovered
2. Your efforts to control the spill — what, when, how. Has a cleanup contractor been contacted?
3. What notice has been given to relevant regulatory agencies?
   a. RCRA requirements of notice and remediation
   b. Local health and safety agencies
4. Are public relations necessary to defuse community concerns? Have you complied with Public Notice requirements?
5. Type and name of insurer
6. Information on your compliance with statutory law on installation, performance standards, and financial responsibility[117]

11.15 GUIDELINES IN REAL ESTATE TRANSACTIONS RELATED TO LUST PROPERTY

Purchase, financing, and leasing of land requires the consideration of an environmental audit, assessment, or investigation. This will raise a "due diligence defense". This may include allocation of environmental risk by agreements. The following guidelines are developed to assist the attorney, landbuyer or seller, and others of some of the elements to be considered in the assessment of risks in a land transaction.

1. Land transaction will require a multidisciplinary exercise of legal, technical, and financial expertise.
2. Renegotiation of terms/price may be required and should be allowed for in the basic documents in the transaction. Clauses requiring future actions/liabilities

by the seller may be appropriate to protect the buyer against prior contamination or failure to inform buyer of problems.Special consideration should be given to the following:

a. Definition of environmental problems
b. Warranties as to conditions, past problems/leaks, prior monitoring, and sampling data/habits
c. Indemnification agreements
d. Termination rights if certain problems/concerns are found on the land by buyer
e. Affirmative and negative covenants
f. Past closing adjustments: these should be backed up with bonds, letters of credit, deed clauses, or other methods to assure that liability of seller survives the contract merger with the deed at time of foreclosure
g. Escrow provisions

3. Lenders must review the transaction to make sure that there is ample money for both repayment of loan and cleanup cost if required, i.e., that the collateral will not be environmentally impaired. Lenders, buyers, and others have a duty to inquire.

4. Environmental Audits — whether at seller's expense or borrower's expense, an environmental audit should be considered. Elements of an environmental audit include the following:

a. A records check of the property to include:
   i. A site history including names of owners and a record of prior use of the site and the length of time property has been used as it is now.
   ii. Lists published by state, county, local, or Federal agencies such as EPA's National Priorities List of Hazardous Sites, RCRA Notifiers List and the Comprehensive Environmental Response Compensation and Liability Information System (CERCLIS) list.
   iii. Court records that show enforcement actions or pending enforcement actions by EPA, state, county, or local authorities for environmental hazards.
b. Observable indications of problems such as
   i. Unusually discolored soil
   ii. Dead vegetation
   iii. Seeping ground liquids
   iv. Discarded trash and other evidence of lax management
   v. History of demolition on or near site
c. Area site maps, drawings, and aerial photos as appropriate
d. Nature of area
   i. Commercial, industrial, residential
   ii. Source of drinking water — well, municipal
e. Comments or complaints from neighbors
f. Site flow of run off
   i. Through property via surface drainage
   ii. Storm drainage system
g. List of fuel and other hazardous substances used at location
h. Inventory tanks
   i. Tested or leaks? Date
   ii. Are underground tanks routinely inventoried?
   iii. Does the site operator currently have a contingency plan?

5. Overall Property Description checklist:

__ Building Specifications	__ Neighborhood Zoning Maps
__ Historical Aerial Photos	__ Neighborhood Land Use Maps
__ Current Aerial Photos	__ List of Commercial Tenants On-Site
__ Title History	__ Verification of Public Water and Sewer
__ Site Survey	__ Interviews with Builder, and/or Property
__ Interviews with Local Fire,	Manager
Health, Land Use or	__ Other _____
Environmental Enforcement	
Officials	

6. Are there any petroleum storage and/or delivery facilities (including gas stations) or chemical manufacturing plants located on adjacent properties?[118]

7. Have UST facilities been maintained in accordance with sound industry standards (e.g., API [American Petroleum Institute] Bulletins 1621 and 1623; NFPA [National Fire Prevention Association] Bulletins 329, 70,77)?[119]

8. Age of tanks
   a. Date of installation
   b. Installer
   c. Type of material
   d. Why replaced

9. Are there any deactivated UST's on the property? Were they deactivated in accordance to sound industry practices such as API Bulletins #1604 and #2202 or NFPA Bulletin #30?

10. Are there any prior reports by an environmental consultant on the property? What were these assessment results?

11. Have lenders refused to lend money on this property?

12. Name of insurance company and policy number of prior environmental pollution coverage.

A buyer/lessee should obtain a certification from the seller/owner relating to use of the property. Certification should include the following:

1. Use of the site including a list of all hazardous chemicals used on site. How and where the hazardous chemicals were disposed. (Future practices may allow for an inventory audit of chemicals/oil/gasoline in and chemicals/oil/gasoline sold/processed).

2. Owner/seller has not received notification from a federal, state, or local government or private plaintiff in regard to pending or threatened superfund or superlien liability, common law suit, or violation of any federal, state, or local ordinance.

3. To the best of the seller/owners knowledge, no environmental hazards have been identified on the subject property, adjacent properties or within the immediate area (1-mile radius) except as disclosed in the sales contract or lease agreement.

4. Any knowledge of use, problems, complaints against the property prior to owner/lessor's acquisition of right in the property.

11.16 FUTURE LITIGATION

The following recent cases are a window of the direction for future litigation for LUST. Including higher standards and stronger application of common law principles, the litigation trend is likely to include more litigation over legislative interpretation and insurance claims and definitions.

In Fischer & Porter Co. v. Liberty Mut. Ins. Co.,[120] Fischer and Porter sought reimbursement by the insurance company for cleaning up contaminated groundwater. Continuous leaking from tanks where management continuously dumped chemicals was not a "sudden and accidental" occurrence within exception to pollution exclusion clauses of the insurance company's general liability policy. More litigation over terms of insurance policy can be expected as firms attempt to shift the cost and burdens of pollution from LUST.

In Rodenbeck v. Marathon Petroleum Company,[121] plaintiffs alleged negligence against defendants for contamination of soil and damages. Plaintiff owners of real estate had previously leased the property to defendant UST owner. Plaintiffs were unable to sell property to a prospective buyer who exercised his option to not purchase when borings showed substantial contamination. The court held that plaintiff's previously signed agreements with defendant to

> ...release the other from *all* [emphasis added] claims and obligations of any character or nature whatsoever arising out of or in connection with said agreement...[122]

operated as a bar for plaintiff to recover for claims of damages under CERCLA or state Underground Storage Tank legislation. The court held that private parties retain the ability to contract cleanup liability.[123]

11.17 CONCLUSION

Success or failure of common law cases often rests upon the individual facts, the presentation of the facts of the case by plaintiff's and defendant's lawyers, and the various states' interpretation of the law. Modern regulation has in part replaced the common law as a major concern for "owners" and "operators" of regulated tanks. However, regulated and nonregulated tank owners are covered by the common law rules, while a defendant who complies or attempts to comply with federal and state UST rules may not use compliance as a shield against common law actions for damages by neighbors adversely impacted by UST leaks. In fact, the failure to comply with UST regulations could be used as the standard of care required of the owner/operator by a common law jury. The case law shows a trend in favor of the victim of LUST.

NOTES

1. See City of Milwaukee v. Illinois and Michigan, 451 U.S. 304 (1981) and International Paper Co. v. Ovelletee 479 U.S. 805 (1987). See also material at Note 65.
2. See Calabresi, G., *The Costs of Accidents: A Legal and Economic Analysis,* Yale Press, New Haven, 1970.
3. 42 U.S.C. Sec. 6901 *et seq.*
4. 94 N.W.2d 799 (Mich 1959).
5. *Id.* 801.
6. New York Telephone Co. v. Mobil Oil Corp., 473 N.Y.S. 2d 172, 174.
7. *Id.*
8. Restatement of the Law Section 352 Comment (a) and Section 353.
9. New York Telephone Co. v. Mobil Oil Corp., 473 N.Y.S. 2d 172, 175 (1984).
10. *Id.*
11. 311 S.E.2d 757 (Va. 1984).
12. *Id.* 760.
13. *Id.*
14. *Id.* 760, 761. See also Exxon Corp. v. Yarema, 516 A.2d 990 (Md. App. 1986).
15. *Id.* 761.
16. 321A 2d 100 (R.I. 1974).
17. But see later case in different jurisdiction: Exxon Corp. v. Yarema, 516 A.2d 990 (Md. App. 1986).
18. *Id.* 102.
19. *Id.* 103.
20. 301 A.2d 705 (Pa. 1973).
21. *Id.* 708.
22. *Id.*
23. Pryor v. Chambersburg Oil & Gas Co., 103 A.2d 425, 427–428 (Pa. 1954).
24. 3 AM JUR Proof of Facts, 3d 532 (1989).
25. South Central Bell Telephone v. Texaco, Inc., 418 So.2d 531, 533 (La. 1982). See also Exxon *op. cit.* 1005.
26. Bean v. Sears, Roebuck and Company, 276 A.2d 613, 616 (Vt. 1971).
27. Machen v. Gulf Oil Corporation, 184 S.2d 550, 556 (La. 1966).
28. Pryor v. Chambersburg Oil and Gas Co., *op. cit.* 22.
29. 476 P.2d 346, 349 (Okla. 1970).
30. 302 A.2d 121 (Conn. 1972).
31. *Id.* 125.
32. City of Douglas, Batly Cty. v. Tri-Co. Fertilizer, Inc., 519 F.2d 724 (Kan. 1974).
33. Branch v. Western Petroleum, Inc., 657 P.2d 267, 276 (Utah 1982).
34. *Id.* 274.
35. 257 A.2d 138, 139 (Md. 1969).
36. 427 N.W.2d 171 (Iowa 1988).
37. *Id.* 175.
38. Akers v. Ashland Oil & Refining Co., 80 S.E.2d 884, 888-889 (W.Va. 1954).
39. Restatement of Torts 2nd Section 520.
40. See 62 AM JUR 2d Premises Liability, Section 6 (1990). Arizona, California, Connecticut, Georgia, Indiana, Kansas, Louisiana, Maryland, Massachusetts, Michigan, Minnesota, Mississippi, Missouri, New Jersey, New Mexico, Ohio, Oregon, South Carolina, Utah, West Virginia, and Wisconsin have adopted a rule of strict

liability for non-natural use. See Keeton et al., *Processor and Keeton on the Law of Torts*, Section 78 (1984).

41. Restatement of Torts 2d Section 520 (1976).
42. *Id.*
43. Yomers, *op. cit.* 34 at 141.
44. *Id.* 140.
45. For a contrary view on similar facts, see Grene v. Spinning, 48 S.W.2d 51 (Kansas 1932). For a similar view, see Berger v. Minneapolis Gaslight Co., 62 N.W.336 (1895). For a modern case, see Exxon *op. cit.,* 13 at 1005.
46. 498 F.2d 619 (1974).
47. Section 13 of the Rivers and Harbors Act of 1899, 33 U.S.C. Section 407.
48. U.S. v. White Fuel Corporation, 498 F. 2d 619, 622 (1974).
49. *Id.* 623. The Court added

> The defendant, if a substantial business enterprise, would usually have exclusive control of both the expertise and the relevant facts; it would be difficult indeed, and to no purpose, for the government to have to take issue with elaborate factual and theoretical arguments concerning who, why and what went wrong. A municipality may require dog owners to keep their dogs off the public streets, and the court may enforce the ordinance by criminal sanctions without paying attention, except in mitigation, to the owner's tales concerning his difficulty in getting Fido to stay home. In the present circumstances we see no unfairness in predicating liability on actual non-compliance rather than either intentions or best efforts. *See* O. W. Holmes, The Common Law 49 (1881). Whatever occasional harshness this could entail is offset by the moderateness of the permitted fine, the fact that the statute's command — to keep refuse out of the public waters — scarcely imposes an impossible burden[8] (FN 8: The Refuse Act makes discharge of refuse illegal only if no permit is obtained. The question whether there are industries which must be allowed to discharge refuse because of their peculiar needs is best left to Congress and the designated agencies. *See* United States v. Kennebec Log Driving Co., 491 F.2d 562, 568 (1st Cir. 1973), cert. denied, — U.S. — , 94 S. Ct. 2607, 41 L.Ed.2d 214 (1974).

50. See Note 112.
51. 607 S.W.2d 677 (Ark. 1980).
52. *Id.* 679.
53. 657 P.2d 267, 273 (Utah 1982).
54. L. R. 1 Ex. 265 (1866), aff'd in Rylands v. Fletcher, L.R. 3 H. L. 330 (1868).
55. Branch *op. cit.* 32 at 274.
56. South Central Bell Tel. Co. v. Hartford ACC, 385 So.2d 830 (La. 1980).
57. *Id.* 833. This is a "theory of negligence without fault."
58. 519 F. Supp. 515, 516 (1981).
59. 566 P.2d 75, 177-178 (OR. 1977).
60. *Id.* The action was brought in trespass and required a finding of negligence or the carry on of an ultra hazardous activity to find a trespass.
61. Phillip v. Sun Oil Co., 121 N.E.2d 249, 251 (N.Y. 1954). See also Henningan v. Atlantic Refining Co. 252 F. Supp. 667 (1966).
62. Dillion v. Acme Oil Co., 210 N.Y.S. 259. See also Restatement of Torts 2d Section 158.
63. Hudson v. Peavey Oil Co., 566 P.2d 175, 177 (OR. 1977).
64. The federal common law encompasses that case law which is judicially fashioned, rather than interpretive of Federal Constitution, statutes, treaties, administrative regulations or executive orders.
65. Sinclair Refining Co. v. Keister, 64 F.2d 537 (1933).

66. See National Audubon Soc. v. Department of Water, 869 F.2d 1196 (CA9. CAL 1988), Middlesex County Sewerage Authority v. Sea Clammers Association, 453 US 1 (1981), Milwaukee v. Illinois, 451 US 304 (1981) holding that the federal law of nuisance in the area of water pollution is entirely preempted by the more comprehensive scope of the CWA (33 USC Sections 1241 et seq.) and United States v. Kin-Bisc. Inc., 532 F Supp 699 (DC NJ, 1982) holding the same under the Clean Air Act (42 USC Sections 7401 et seq.)

67. 257 A.2d 138, 141, 142 (Md 1969).

68. *Id.* 142.

69. 452 A.2d 586, 588 (Pa. 1982).

70. *Id.* 590.

71. SOCONY Mobil Co., Inc. v. Southwestern Bell Tel. Co., 518 S.W.2d 257, 267 (Texas 1974).

72. Cooper v. Whiting Oil Company, Incorporated, 311 S.E.2d 757, 760-761 (Va 1984).

73. Sinclair Refining Co. v. Bennett, 123 F.2d 884, 886 (1941). See Continental Oil Co. v. Berry, 52 S.W.2d 953 (Texas 1932).

74. Monroe "66" Oil Company v. Hightower, 180 So.2d 3, 10 (La. 1965).

75. 480 A.2d 1012, 1019 (Pa. Super. 1984).

76. *Id.* 1020.

77. *Id.*

78. *Id.* 1020-1034.

79. Ewell v. Petro Processors of Louisiana, Inc., 364 So.2d 604, 607 (1978).

80. *Id.* 607-608.

81. 364 So. 2d 604, 609 (1978).

82. *Id.*

83. *Id.* ($5000 was also awarded for mental anguish.) Pre-CERCLA and RCRA, the moral of this case is to buy the affected property. Post CERCLA and RCRA, the property would need to be cleaned up.

84. 276 A.2d 613, 616 (Vt 1971).

85. The words of Judge Learned Hand at 617.

86. Bousquet v. Commonwealth, 372 N.E.2d 257, 258 (Mass. 1978).

87. *Id.* Restatement of Agency 2nd Sections 214, 220(2) 251(a).

88. 516 A.2d 990, 993 (Md. App. 1986).

89. *Id.* 1007.

90. *Id.* 1009.

91. Searcy, M., et al., A Guide to Toxic Torts, Ch. 10, Proof of Causation, New York (1987).

92. Com. of Puerto Rico v. SS Zoe Colocotroni, 628 F.2d 652, 673-674 (1980). See also CWA 33 U.S.C. Section 1321 (f)(4) (1991) authorizing restoration or replacement of natural resources.

93. *Id.*

94. 372 N.E.2d 257, 258 (Mass. 1978).

95. Hazardous and Solid Waste Amendments of 1984, 42 U.S.C.A. Sections 6901 et seq. (1991) and RCRA 42 U.S.C.A. Sections 6921 et seq. (1991).

96. See for example Fleck v. Timmons, 543 A.2d 148, 151-152 (Pa. Super. 1988) where the court held that Pennsylvania's Solid Waste Management Act (SWMA) restricts the right to seek relief for the violation of its provisions to the Department of Environmental Resources. The Court held that legislature did not afford the

private citizen the right to use the SWMA to institute a private cause of action. The court went on to state that the

> ...legislature did not intend to allow...legal presumptions created in...SWMA to be used in any legal proceedings other than a proceeding instituted by the [State]...To hold otherwise would be to allow private citizens to recover for hazardous waste activities with no proof of fault, negligence, or causation.

97. See cases at Notes 1 and 65.
98. "Violation of the law cannot be considered reasonable conduct," O'Brien v. City of O'Fallon, 400 N.E.2d 456, 462 (1980).
99. 42 U.S.C. Sections 6991b.(C) (1988).
100. Miller v. Schoene, 276 U.S. 272, 279 (1928).
101. *Id.* Penn Central Transportation Co. v. New York, 438 U.S. 104 (1978) and Euclid v. Ambler Realty Co., 272 U.S. 365 (1926).
102. 765 P.2d 933 (Wyo. 1988).
103. Mower v. Ashland Oil & Refining Co. Inc., 518 F.2d 659 (1975).
104. Pryor *op. cit.* 22 at 248.
105. *Id.*
106. South Central Bell Telephone v. Texaco, Inc., 418 So.2d 531 (La. 1982).
107. *Id.* 533.
108. 473 N.Y.S.2d 172, 173-174.
109. Individual state statutes vary as to time the action must be initiated. For example, Virginia's is 2 years for tort action but North Carolina's is 3 years.
110. Wilson v. McLeod Oil Co. Inc. 586, 596 (N.C. 1990).
111. *Id.*
112. *Id.* Although the state act did not contain a statute of limitations, the court applied the general state statute of limitations. The court implied a right of action under state "Oil Pollution...Act", a result contrary to other states. See material at Notes 95 and 96.
113. *Id.* 623, 624.
114 U.S. Environmental Agency, More About Leaking Underground Storage Tanks: A Background Booklet for the Chemical Advisory 6 (Oct. 1984). See also material in this chapter infra notes.
115. See Exxon *op. cit.* for a case where punitive damages were assessed for delay.
116. The following federal statutes have notification requirements:
> *Clean Water Act*, 33 USC. Section 1321(b)(3) and 40 CFR. Section 110.3 require notification of any discharge of oil that reaches navigable waters in a quantity that is sufficient to cause a sheen; "Superfund" or *CERCLA*, 42 USC. Section 9603(a) and 40 CFR. Section 302 require notification of a discharge or release into the environment of a reportable quantity of hazardous substances; *SARA Title III*, 42 U.S.C. Section 11004 requires notification to designated local and state organizations of a discharge or release into the environment of a reportable quantity of hazardous substances; and *RCRA*, 42 USC. Sections 6934 and 6991 impose an obligation to notify upon a finding that the release of a hazardous waste may present a substantial hazard to human health or the environment; the EPA can additionally order monitoring, analysis or testing. RCRA also requires one time notice by most UST owners to state agencies. State and local statutes may also require notification.

117. See Chapter 10.
118. Environmental Hazards Management Procedures for Multifamily, Federal National Mortgage Association, p. 78, Aug. 1, 1988. Revised Sept. 1, 1991.
119. *Id.*
120. 656 F. Supp. 132 (E.D. Pa. 1986).

121. 742 F. Supp. 1448.
122. *Id.* 1457. The contract was not one of adhesion even though the bargaining power may have been uneven.
123. *Id.* 1456.

EPA Regional UST Program Offices

(Source: EPA Office of Underground Storage Tanks, Washington, D.C.)

EPA Region 1
(Maine, Vermont, New Hampshire, Massachusetts, Connecticut,
Rhode Island)

U.S. EPA
JFK Federal Bldg.
Mailcode: HPU-1
Boston, MA 02203-2211
617-573-9604

EPA Region 2
(New York, New Jersey, Puerto Rico, Virgin Islands)

U.S. EPA
Hazardous Waste Programs Branch
26 Federal Plaza
Mailcode: 2AWM-HWPB
New York, NY 10278
212-264-1369

EPA Region 3
(Pennsylvania, Virginia, West Virginia, Maryland, District of Columbia,
Delaware)

U.S. EPA
841 Chestnut Bldg.
Mailcode: 3HW31
Philadelphia, PA 19107
215-597-7354

EPA Region 4
(Kentucky, Tennessee, North Carolina, South Carolina, Georgia, Alabama,
Mississippi, Florida)

U.S. EPA
345 Courtland St. N.E.
Mailcode: 4WM-GP
Atlanta, GA 30365
404-347-3866

EPA Region 5
(Wisconsin, Minnesota, Illinois, Indiana, Michigan, Ohio)

U.S. EPA
230 S. Dearborn St.
Mailcode: 5HR-JCK-13
Chicago, IL 60604
312-886-6159

EPA Region 6
(New Mexico, Texas, Oklahoma, Arkansas, Louisiana)

U.S. EPA
1445 Ross Avenue
Mailcode: 6H-A
Dallas, TX 75202-2733
214-655-6755

EPA Region 7
(Iowa, Missouri, Nebraska, Kansas)

U.S. EPA
RCRA Branch
726 Minnesota Avenue
Kansas City, KS 66101
913-236-2852

EPA Region 8
(Montana, Wyoming, Utah, Colorado)

U.S. EPA
999 18th St.
Mailcode: 8-HWM-RM
Denver, CO 80202-2405
303-293-1489

EPA Region 9
(California, Nevada, Arizona, Hawaii, Guam, American Samoa)

U.S. EPA
215 Fremont St.
Mailcode: T-2-7
San Francisco, CA 94105
415-974-8160

EPA Region 10
(Washington, Oregon, Idaho, Alaska)

1200 Sixth Ave.
Mailcode: WD-139
Seattle, WA 98101
206-442-0344

APPENDIX B

State UST Program Offices

(Source: EPA Office of Underground Storage Tanks, Washington, D.C.)

Alaska
Dept. of Environmental Conservation
P.O. Box O
3220 Hospital Drive
Juneau, AK 99811-1800
907-465-2630

Alabama
AL Dept. of Environmental Management
Ground-Water Section/Water Division
1751 Congressman W. Dickerson Dr.
Montgomery, AL 36130
205-271-7832

Arkansas
AR Dept. of Pollution Control & Ecology
P.O. Box 9583
8001 National Dr.
Little Rock, AR 72219
501-562-7444

Arizona
AZ Dept. of Environmental Quality
2005 N. Central Ave.
Phoenix, AZ 85004
602-257-6984

California
CA State Water Resources Control Board
P.O. Box 944212
214 T St.
Sacramento, CA 94244-2120
916-739-2421

Colorado
CO Dept. of Health
Hazardous Materials and Waste Management Program
UST Program
4210 East 11th Ave.
Denver, CO 80220
303-331-4830

Connecticut
CT Dept. of Environmental Protection
Hazardous Material Management Unit/UST Program
State Office Bldg.
165 Capitol Ave.
Hartford, CT 06106
203-566-4630

District of Columbia
DC Dept. of Consumer and Regulatory Affairs
Pesticides & Hazardous Waste Management Branch
614 H St. N.W., Rm. 505
Washington, D.C. 20001
202-783-3205

Delaware
DE Dept. of Natural Resources and Environmental Control
Division of Air & Waste Management
715 Grantham Lane
New Castle, DE 19720
302-323-4588

Florida
FL Dept. of Environmental Regulation
Tank Section
Twin Towers Ofc. Bldg. Rm 202
2600 Blair Stone Rd.
Tallahassee, FL 32399-2400
904-488-3936

Georgia
GA Environmental Protection Division
UST Unit
3420 Norman Berry Dr., 7th Floor
Hapeville, GA 30354
404-669-3927

Hawaii
HI Dept. of Health
645 Halekauwile St., 2nd Floor
Honolulu, HI 96813
808-548-8837

Iowa
IA Dept. of Natural Resources
Henry A. Wallace Bldg.
900 East Grand
Des Moines, IA 50319
515-281-8692

Idaho
ID Dept. of Health & Welfare
450 West State St.
Boise, ID 83720
208-334-5847

Illinois
IL Office of State Fire Marshal
Division of Petroleum & Chemical Safety
1035 Stevenson Dr.
Springfield, IL 62703-4259
217-785-5878

Indiana
IN Dept. of Environmental Management
UST Program
105 South Meridian St.
Indianapolis, IN 46225
317-243-5055

Kansas
KS Dept. of Health & Environment
Forbes Field, Bldg. 740
Topeka, KS 66620
913-296-1597

Kentucky
KY Dept. of Environmental Protection
Hazardous Waste Branch
18 Reilly Rd.
Frankfort, KY 40601
502-564-6716

Louisiana
Dept. of Environmental Quality
P.O. Box 44274
438 Main St.
Baton Rouge, LA 70804
504-342-7808

Massachusetts
MA Dept. of Public Safety
PO Box 490
East Street, Bldg. #5
Tewksbury, MA 01876
508-851-9813

Maryland
MD Dept. of Environment
2500 Broening Highway
Baltimore, MD 21224
301-631-3442

Maine
Dept. of Environmental Protection
Bureau of Hazardous Materials and Solid Waste Control
State House Station #17
Augusta, ME 04333
207-289-2651

Michigan
MI Dept. of State Police
Fire Marshall Division
7150 Harris Dr.
Lansing, MI 48913
517-322-5470
MI residents: 1-800-MICH UST

Minnesota
MN Pollution Control Agency
UST Program
520 Lafayette Road
St. Paul, MN 55155
612-296-7743

Missouri
MO Dept. of Natural Resources
P.O. Box 176, 205 Jefferson Street
Jefferson City, MO 65102
314-751-7428

Mississippi
Dept. of Natural Resources
Bureau of Pollution Control
UST Section
P.O. Box 10385, 2380 Hwy 80 West
Jackson, MS 39289-0385
601-961-5171

Montana
Dept. of Health & Environmental Science
Solid & Hazardous Waste Bureau
Cogswell Bldg., Room B-201
Helena, MT 59620
406-444-5970

North Carolina
NC Dept. of Natural Resources & Community Dev.
Division of Environmental Management
Ground-Water Operations Branch
P.O. Box 27687, 512 N. Salisbury
Raleigh, NC 27611-7687
919-733-7015

North Dakota
ND Dept. of Health
Division of Waste Management
Box 5520, 1200 Missouri Ave.
Bismarck, ND 58502-5520
701-224-2366

Nebraska
NE State Fire Marshal
P.O. Box 94677
246 South 14th
Lincoln, NE 68509-4677
402-471-9465

New Hampshire
NH Dept. of Environmental Services
Waste Supply & Pollution Control Division
6 Hazen Dr.
P.O. Box 95
Concord, NH 03301
603-271-3444

New Jersey
NJ Dept. of Environmental Protection
Division of Water Resources (CN-029)
401 East State Street
Trenton, NJ 08625
609-984-3156

New Mexico
NM Environmental Improvement Division
UST Bureau
1190 St. Francis Drive
Harold Runnels Bldg.
Santa Fe, NM 87504
505-827-0188

Nevada
NV Dept. of Conservation & Natural Resources
Division of Environmental Protection
Capitol Complex
201 S. Fall St.
Carson City, NV 89710
702-885-5872

New York
NY Dept. of Environmental Conservation
Bulk Storage Section, Division of Water
50 Wolf Road, Room 326
Albany, NY 12233-3520
518-457-4351

Ohio
OH Dept. of Commerce
Division of Fire Marshall
Bureau of UST Regulations
7510 East Main Street
P.O. Box 525
Reynoldsburg, OH 43068
614-752-8200

Oklahoma
OK Corporation Commission
UST Program
Jim Thorpe Bldg.
2101 N. Lincoln Blvd.
Oklahoma City, OK 73105
405-521-3107

Oregon
OR Dept. of Environmental Quality
811 SW Sixth Ave
Portland, OR 97204
503-229-6652

Pennsylvania
PA Dept. of Environmental Resources
Bureau of Water Quality Management
Non-point Source & Storage Tank Section
Fulton Building, 12th Floor
P.O. Box 2063
Harrisburg, PA 17120
717-787-8184

Rhode Island
RI Dept. of Environmental Management
291 Promenade St.
Providence, RI 02908
401-277-2234

South Carolina
SC Dept. of Health and Environmental Control
Ground-Water Protection Division
2600 Bull St.
Columbia, SC 29201
803-734-5332

South Dakota
SD Dept. of Water & Natural Resources
Office of Water Quality
523 East Capitol
Joe Foss Bldg.
Pierre, SD 57501-3181
605-773-3351

Tennessee
TN Dept. of Health & Environment
Division of Superfund/UST Program
706 Church St.
Doctors Bldg., 2nd Floor
Nashville, TN 37219-5404
615-741-4094

Texas
Texas Water Commission
UST Section
P.O. Box 13087
1700 North Congress
Austin, TX 78711
512-463-7786

Utah
UT State Health Dept.
Bureau of Solid & Hazardous Waste
288 N. 1460 West
Salt Lake City, UT 84116-0700
801-538-6752

Virginia
VA State Water Control Board
P.O. Box 11143
2111 N. Hamilton St.
Richmond, VA 23230-1143
804-367-6685

Vermont
VT Dept. of Environmental Conservation
103 South Main St.
Waterbury, VT 05676
802-244-8702

Washington
WA Dept. of Ecology
Solid & Hazardous Waste Program/UST Unit
4224 Sixth Ave.
Rowesix, Bldg. 4
Mailstop PV-11
Olympia, WA 98504-8711

Wisconsin
WI Dept. of Industry, Labor, & Human Relations
P.O. Box 7969
201 East Washington Ave.
Madison, WI 53707
608-267-9725

West Virginia
WV Dept. of Natural Resources
Division of Waste Management
1201 Greenbriar Street
Charleston, WV 25311

Wyoming
WY Dept. of Environmental Quality
Water Quality Division
Herschler Bldg, 4th Floor West
122 W. 25th St.
Cheyenne, WY 82002
307-777-7085

American Samoa
EPA
Office of the Governor
American Samoa Government
ATTN: UST Program
Pago Pago, American Samoa 96799
684-633-2682

Commonwealth of Northern Mariana Islands
CNMI Division of Environmental Quality
PO Box 1304
Commonwealth of Northern Mariana Islands
Saipan, CM 96950
607-234-6984

Guam
EPA
IT&E
Harmon Plaza, Complex Unit D-107
130 Rojas Street
Harmon, Guam 96911
671-646-8863

Puerto Rico
PR Water Quality Control
Environmental Quality Board
P.O. Box 11488
Commonwealth of Puerto Rico
Santurce, PR 00910
809-725-8410

Virgin Islands
VI Environmental Protection Division
Dept. of Planning and National Research
Suite 213, Nisky Center
Charlotte Amalie
St. Thomas, VI 00802
809-774-3320

APPENDIX C

Industry Information Sources

Useful industry information can be obtained from the following organizations:

AMERICAN PETROLEUM INSTITUTE (API)
1220 L Street, N.W.
Washington, D.C. 20005
202-682-8000

NATIONAL FIRE PROTECTION ASSOCIATION (NFPA)
Batterymarch Park
Quincy, MA 02269
617-770-3000

PETROLEUM EQUIPMENT INSTITUTE (PEI)
P.O. Box 2380
Tulsa, OK 74101
918-743-9941

STEEL TANK INSTITUTE
P.O. Box 4020
Northbrook, IL 60065
312-498-1980

Index

abandonment, procedures, 32
abnormally dangerous activity, 253–255
accidental release, definition, 151
affordability, 186
Alabama, soil cleanup standards in, 70
aliphatics, 61
alkane, 61
alternate financial responsibility
    mechanisms, 156
American Petroleum Institute (API), 20
  methods recommended for removal of
    vapor, 35
  Publication 1604, 20, 30, 31
  Publication 1631, 20
  Publication 2015, 20, 31
analytical limitations, 70
analytical methods, selection of, 60
annual aggregate coverage, 120, 121,
    134, 145, 146, 162
API, see American Petroleum Institute
aromatics, 61
aromatic volatile organics, 65
arsenic, 69
asphalts, 61, 62
audits, 50
availability, 186

background soil concentrations, 70
bankruptcy, 159
banks, environmental assessments
    required by, 71
benzene, toluene, ethylbenzene, xylene
    (BTEX), 61, 62, 66, 68, 69
bid package, 93
bioremediation, 47
BOCA, see Building Officials and Code
    Administration
bodily injury, definition, 151
bond rating test, 148
BTEX, see benzene, toluene,
    ethylbenzene, xylene

Building Officials and Code Administra-
    tion (BOCA), 25
buyer/lessee, 266

Category I firm, 122, 123, 161
Category II firm, 122, 123, 125, 161
Category III firm, 122, 124, 127, 128,
    161
Category IV firm, 123, 124, 127, 128,
    130, 161
causation, 257, 259
CERCLA, see Comprehensive Environ-
    mental Response, Compen-
    sation, and Liability Act
CFR, see Code of Federal Regulations
CGI, see combustible gas indicator
change in service, 19
circumstantial evidence, 257
citizen suits, 259
claims made, 181
  insurance, 151, 152
  policies, 176
claims management, 215, 229
cleanup standards, 47
closure
  industry standards, 20
  procedures, 20
  records, 21
  sequence of tasks, 9
  timeline, 9
Code of Federal Regulations (CFR), 13
colorimetric tubes, 74, 75
combustible gas indicator (CGI), 35
commercial cleaning facilities, 38
common law, 247, 248, 267
common law trespass, 256
compliance dates for financial responsi-
    bility, 122, 123, 135,
    143–145, 161
compliance deadlines, 114
composite sampling, 58